Science, Technology and Medicine in Modern History

General Editor: John V. Pickstone, Centre for the History of Science, Technology and Medicine, University of Manchester, England (www.man.ac.uk/CHSTM)

One purpose of historical writing is to illuminate the present. At the start of the third millennium, science, technology and medicine are enormously important, yet their development is little studied.

The reasons for this failure are as obvious as they are regrettable. Education in many countries, not least in Britain, draws deep divisions between the sciences and the humanities. Men and women who have been trained in science have too often been trained away from history, or from any sustained reflection on how societies work. Those educated in historical or social studies have usually learned so little of science that they remain thereafter suspicious, overawed, or both.

Such a diagnosis is by no means novel, nor is it particularly original to suggest that good historical studies of science may be peculiarly important for understanding our present. Indeed this series could be seen as extending research undertaken over the last half-century. But much of that work has treated science, technology and medicine separately; this series aims to draw them together, partly because the three activities have become ever more intertwined. This breadth of focus and the stress on the relationships of knowledge and practice are particularly appropriate in a series which will concentrate on modern history and on industrial societies. Furthermore, while much of the existing historical scholarship is on American topics, this series aims to be international, encouraging studies on European material. The intention is to present science, technology and medicine as aspects of modern culture, analysing their economic, social and political aspects, but not neglecting the expert content which tends to distance them from other aspects of history. The books will investigate the uses and consequences of technical knowledge, and how it was shaped within particular economic, social and political structures.

Such analyses should contribute to discussions of present dilemmas and to assessments of policy. 'Science' no longer appears to us as a triumphant agent of Enlightenment, breaking the shackles of tradition, enabling command over nature. But neither is it to be seen as merely oppressive and dangerous. Judgement requires information and careful analysis, just as intelligent policy-making requires a community of discourse between men and women trained in technical specialities and those who are not.

This series is intended to supply analysis and to stimulate debate. Opinions will vary between authors; we claim only that the books are based on searching historical study of topics which are important, not least because they cut across conventional academic boundaries. They should appeal not just to historians, nor just to scientists, engineers and doctors, but to all who share the view that science, technology and medicine are far too important to be left out of history.

Titles include:

Julie Anderson, Francis Neary and John V. Pickstone
SURGEONS, MANUFACTURERS AND PATIENTS

A Transatlantic History of Total Hip Replacement
Roberta E. Bivins
ACUPUNCTURE, EXPERTISE AND CROSS-CULTURAL MEDICINE

Linda Bryder
WOMEN'S BODIES AND MEDICAL SCIENCE
An Inquiry into Cervical Cancer

Roger Cooter
SURGERY AND SOCIETY IN PEACE AND WAR
Orthopaedics and the Organization of Modern Medicine, 1880–1948

Catherine Cox, Hilary Marland
MIGRATION, HEALTH AND ETHNICITY IN THE MODERN WORLD

Jean-Paul Gaudillière and Ilana Löwy (editors)
THE INVISIBLE INDUSTRIALIST
Manufacture and the Construction of Scientific Knowledge

Jean-Paul Gaudillière and Volker Hess (editors)
WAYS OF REGULATING DRUGS IN THE 19TH AND 20TH CENTURIES

Christoph Gradmann and Jonathan Simon (editors)
EVALUATING AND STANDARDIZING THERAPEUTIC AGENTS, 1890–1950

Sarah G. Mars
THE POLITICS OF ADDICTION
Medical Conflict and Drug Dependence in England since the 1960s

Alex Mold and Virginia Berridge
VOLUNTARY ACTION AND ILLEGAL DRUGS
Health and Society in Britain since the 1960s

Ayesha Nathoo
HEARTS EXPOSED
Transplants and the Media in 1960s Britain

Neil Pemberton and Michael Worboys
MAD DOGS AND ENGLISHMEN (hardback 2007)
Rabies in Britain, 1830–2000

Neil Pemberton and Michael Worboys
RABIES IN BRITAIN (paperback 2012)
Dogs, Disease and Culture, 1830–1900

Cay-Rüdiger Prüll, Andreas-Holger Maehle and Robert Francis Halliwell
A SHORT HISTORY OF THE DRUG RECEPTOR CONCEPT

Thomas Schlich
Surgery, Science and Industry
A Revolution in Fracture Care, 1950s–1990s

Eve Seguin (*editor*)
INFECTIOUS PROCESSES
Knowledge, Discourse and the Politics of Prions

Crosbie Smith and Jon Agar (*editors*)
MAKING SPACE FOR SCIENCE
Territorial Themes in the Shaping of Knowledge

Stephanie J. Snow
OPERATIONS WITHOUT PAIN
The Practice and Science of Anaesthesia in Victorian Britain

Carsten Timmermann
A HISTORY OF LUNG CANCER
The Recalcitrant Disease

Carsten Timmermann and Julie Anderson (*editors*)
DEVICES AND DESIGNS
Medical Technologies in Historical Perspective

Carsten Timmermann and Elizabeth Toon (*editors*)
CANCER PATIENTS, CANCER PATHWAYS
Historical and Sociological Perspectives

Jonathan Toms
MENTAL HYGIENE AND PSYCHIATRY IN MODERN BRITAIN

Duncan Wilson
TISSUE CULTURE IN SCIENCE AND SOCIETY
The Public Life of a Biological Technique in Twentieth Century Britain

---

Science, Technology and Medicine in Modern History
Series Standing Order ISBN 978-0-333-71492-8 hardcover
Series Standing Order ISBN 978-0-333-80340-0 paperback
(*outside North America only*)

You can receive future titles in this series as they are published by placing a standing order. Please contact your bookseller or, in case of difficulty, write to us at the address below with your name and address, the title of the series and one of the ISBNs quoted above.

Customer Services Department, Macmillan Distribution Ltd, Houndmills, Basingstoke, Hampshire RG21 6XS, England

A History of Lung Cancer

The Recalcitrant Disease

Carsten Timmermann
Centre for the History of Science, Technology and Medicine (CHSTM),
University of Manchester, UK

palgrave
macmillan

First published 2014 by
PALGRAVE MACMILLAN

Palgrave Macmillan in the UK is an imprint of Macmillan Publishers Limited, registered in England, company number 785998, of Houndmills, Basingstoke, Hampshire RG21 6XS.

Palgrave Macmillan in the US is a division of St Martin's Press LLC, 175 Fifth Avenue, New York, NY 10010.

Palgrave Macmillan is the global academic imprint of the above companies and has companies and representatives throughout the world.

Palgrave® and Macmillan® are registered trademarks in the United States, the United Kingdom, Europe and other countries

ISBN: 978-1-4039-8802-7

This book is printed on paper suitable for recycling and made from fully managed and sustained forest sources. Logging, pulping and manufacturing processes are expected to conform to the environmental regulations of the country of origin.

A catalogue record for this book is available from the British Library.

A catalog record for this book is available from the Library of Congress.

Contents

List of Figures and Table   vi

Acknowledgements   vii

List of Abbreviations   ix

1   Introduction: The History of a Recalcitrant Disease   1

2   Lung Cancer and Consumption in the Nineteenth   11
    Century: Bodies, Tissues, Cells and the Making of
    a Rare Disease

3   Lungs in the Operating Theatre, circa 1900 to 1950   34

4   Science, Medicine and Politics: Lung Cancer and Smoking,   64
    circa 1945 to 1965

5   Trials and Tribulations: Lung Cancer Treatment, circa 1950   93
    to 1970

6   More Enthusiasm, Please: Preventing, Screening, Treating,   118
    Classifying, circa 1960 to 1990

7   The Management of Stigma: Lung Cancer and Charity,   147
    circa 1990 to 2000

8   Still Recalcitrant? Some Conclusions   171

Notes   177

Bibliography   205

Index   232

List of Figures and Table

Figures

2.1 Stethoscope and Lungs. Illustration from Laennec's 15
De l'auscultation mediate

2.2 Lung cancer cases listed in Adler's book with 24
information on whether or not a microscopic
examination was performed

3.1 Sauerbruch's Negative Pressure Operating Chamber 37

3.2 Morriston Davies's Positive Pressure Anaesthetic Machine 38

4.1 Death rates from cancer, tuberculosis and bronchitis, 68
1916–1959

4.2 Death rates for cancer of the respiratory system 69
compared with rates for all other respiratory diseases

Table

4.1 Participants in the informal MRC lung cancer conference 74
held on 6 February 1947

Acknowledgements

The author would like to thank the Wellcome Trust for their support of the Constructing Cancers project (Programme Grant 068397), which funded the research for this volume. Special thanks are due to John Pickstone for leading the programme, for his many insights, and his 'pruning' of the manuscript; and to Ilana Löwy, who also read the whole manuscript and made valuable suggestions (all remaining mistakes are my own). Many thanks to the other members of our Constructing Cancers team: Elizabeth Toon, Emm Barnes (now Royal Holloway, University of London), Helen Valier (now University of Houston) and Jo Baines, for the fruitful conversations that helped me to better understand the commonalities and differences between lung cancer and other malignant diseases. I am also grateful to the scholars who joined us in our informal 'cancer history' network, for their many stimulating discussions in Manchester, Paris, Bethesda and elsewhere.

I have greatly appreciated the critiques, input and support given by our other colleagues at Manchester's Centre for the History of Science, Technology and Medicine (CHSTM); its Director, Mick Worboys, has been extremely supportive throughout the project. Outside CHSTM, I am grateful to Roger Abbey Smith, Ray Donnelly, Peter Goldstraw, Nick Thatcher and Tom Treasure, for sharing their valuable insights to the history of lung cancer with me, and to the librarians and archivists at the UK National Archives, the Royal College of Surgeons Archives, the Royal London Hospital Archives, the University of Manchester Library and the Wellcome Library for their help. Finally I would like to thank Aya, Jota and Shogo for their patience and good cheer throughout this book's far too long gestation period.

The author and the publishers wish to thank the following for permission to reproduce copyright material:
- Wellcome Images for 'Taking Chest X-ray, 1940', Wellcome Images N0020373, and 'Stethoscope and Lungs', from Laennec, *De l'auscultation mediate*, Wellcome Images L0000603.
- BMJ Publishing Group for the illustration 'Positive Pressure Anaesthetic Machine', from Hugh Morriston Davies, 'The Mechanical Control of Pneumothorax During Operations on the Chest, with a Description of a New Apparatus'. *British Medical Journal* ii (8 July 1911): 61–66; and the graph 'Crude annual death

rates for cancer of respiratory system compared with similar rates for all other respiratory diseases', from David W. Smithers, 'Facts and Fancies About Cancer of the Lung'. *British Medical Journal* i, no. 4822 (6 June 1953): 1235–1239.

– The Royal College of Physicians for the graph 'Standardised death rates from cancer, tuberculosis and bronchitis', from Royal College of Physicians. *Smoking and Health: A Report of The Royal College of Physicians on Smoking in Relation to Cancer of the Lung and Other Diseases*. London: Pitman Medical Publishing, 1962.

List of Abbreviations

AATS	American Association for Thoracic Surgery
ACS	American Cancer Society
AJC	American Joint Committee on Cancer Staging and End Results Reporting
AJCC	American Joint Committee on Cancer
A.P.	Artificial Pneumothorax
ASH	Action on Smoking and Health
BECC	British Empire Cancer Campaign
BLF	British Lung Foundation
BMA	British Medical Association
BMJ	British Medical Journal
CCCR	Co-ordinating Committee on Cancer Research
CNS	Central Nervous System
CRC	Cancer Research Campaign
CT	Computed Tomography
CTO	Cancer Trials Office
CTU	Clinical Trials Unit
ELCAP	Early Lung Cancer Action Project
EMS	Emergency Medical Service
GRO	General Register Office
HIP	Health Insurance Plan
IASLC	International Association for the Study of Lung Cancer
ICRF	Imperial Cancer Research Fund
LSHTM	London School of Hygiene and Tropical Medicine
MPI	Maudsley Personality Inventory
MRC	Medical Research Council (UK)
NCI	National Cancer Institute (US)
NCRI	National Cancer Research Institute (UK)
NHS	National Health Service
NIH	National Institutes of Health (US)
NSCLC	Non-Small Cell Lung Cancer
RCP	Royal College of Physicians
RCS	Royal College of Surgeons
RCT	Randomized Controlled Trial
SCLC	Small Cell Lung Cancer
TNM	Tumour, Node, Metastasis

UCH	University College Hospital
UCL	University College London
UICC	Union Internationale Contre le Cancer (International Union against Cancer)
UK NA	National Archives (UK)
WHO	World Health Organization
YLL	Years of Life Lost

1
Introduction: The History of a Recalcitrant Disease

Mary Benbow was 50 years old when she was admitted to Guy's Hospital in London on 19 August 1841, under the care of Dr Richard Bright, one of the city's leading consultants, and his assistant physician Dr Henry Marshall Hughes, who was known both for his eccentric dress sense and his interest in chest conditions.[1] Hughes recorded Mrs Benbow's case for the *Guy's Hospital Reports*, and what we know about the final months of her life, we know from his report.[2]

Mrs Benbow had been a remarkably healthy person until about two years before her admission to Guy's, when what she thought was a nasty cold confined her to her bed for two months. She was a hard-working and temperate woman; as the wife of a soldier, she had followed the army until she found employment as a washerwoman. Following her cold, she had occasionally coughed up blood. Eventually she turned to the Kingston surgeon Mr Edward Kingsford, who bled her on two occasions and also used other traditional remedies, such as acids, acetate of lead, digitalis, and saline purgatives. But her condition failed to improve and nine months later Kingsford referred her to Guy's Hospital, where he had trained and continued to have contacts. When admitted to the hospital, she looked rather pale but was not in pain, nor was she particularly emaciated. She complained about cough and shortness of breath, and Hughes examined her carefully. He found that the glands in her right armpit and under the collar bone were swollen and noted that the veins directly under the skin on the right side of her abdomen and the lower part of the chest were protruding. The ribs on her right side scarcely moved when she was breathing. When Hughes used his fingers to drum on her chest (a diagnostic procedure known as percussion), there was hardly any sound on the right side. A healthy, air-filled lung would have sounded different. Neither could he hear the typical respiratory murmur through the stethoscope; instead, he noted a 'little coarse mucous rattle'.[3]

Mrs Benbow's symptoms became gradually worse. It became increasingly difficult for her to breathe, and her legs were more and more swollen. About two months after her admission she died, according to Hughes, 'without any particular suffering, or very great emaciation, ... exhausted'.[4] There was not much the doctors at Guy's had felt they could do for Mary Benbow; the illness had taken its natural cause, with only palliative treatments (Hughes used this term). As Guy's was a teaching hospital with its own medical school with physicians interested in research, her body was subjected to a post-mortem examination in order to find out what disease had caused her symptoms. They found that the entire upper part of her right lung had become, in Hughes' words, 'a mass of medullary fungus'. There was no fungus in the usual sense, but the mass looked fungal; and when it was pressed, a creamy fluid and a soft, somewhat brain-like matter exuded, which was why Hughes chose the attribute 'medullary', following the influential French physician R. T. H. Laennec (more about this in Chapter 2). The middle and lower lobes also contained masses of this 'malignant growth', which Hughes, again, described as 'fungoid matter'. He informed Mrs Benbow's doctor that she had been suffering 'from some malignant affection of the lung'.[5] Had she not been referred to Guy's and exposed to the new diagnostic techniques employed by doctors there, and had her body not been subjected to a post-mortem examination, it is likely that Mary Benbow's illness would have been viewed as a sad but fairly common case of 'consumption', a set of symptoms also called 'phthisis' and then increasingly identified as tuberculosis. As she became one of Bright's and Hughes's patients, however, hers is one of the few cases of lung cancer that made it into the medical literature before 1900. Meanwhile, from the 1850s, cancer (in general) had turned from a disease of abnormal tissues into a disease of abnormal cells.

In the early twentieth century, a growing number of cases of lung cancer were detected, leading to debates as to whether this disease was in fact becoming more common, or whether the increase in incidence was an artefact of better means of diagnosis. A steep increase in diagnosed lung tumours in the 1940s convinced most observers that this was no artefact, and the international debate turned to possible causes, homing in on tobacco consumption and air pollution. By the late 1950s most experts agreed that cigarettes were to blame: they had been mass-produced since the 1880s – and clearly not the cause of Mary

Benbow's cancer. As epidemiologists were busy pinning down the main cause of lung cancer, the practitioners treating lung cancer patients were no longer as helpless as those treating Mary Benbow. From the 1940s, surgery was turning into a standard treatment, at least in better equipped hospitals, especially those with chest units. Finally there was something that could be done. One of the patients being subjected to lung cancer surgery was King George VI in 1951 (more on this in Chapter 5). But few lives were significantly prolonged by the intervention – even the king died four months after his operation. Other treatment modalities – namely radiotherapy developed between the wars, and the later methods of cancer chemotherapy – failed to make much of a difference.

By the 1990s, surgery was still the standard treatment, with other modalities employed to palliate, or when the tumour had recurred and it was felt that 'something had to be done'. This was the pattern, even if none of the medical interventions was likely to prolong a patient's life, let alone improve its quality during the final months or weeks. Consider Frank Craig, for example, the husband of the British religious broadcaster and author Mary Craig, who was diagnosed with a lung tumour in 1991.[6] The Craigs had suspected that something was seriously wrong with him when he coughed up blood a few months earlier. A chest X-ray revealed a shadow on the lung and he underwent a procedure known as a bronchoscopy: a flexible tube was inserted which allowed doctors to see into his bronchus and which could also be used to remove a tissue sample. The examination of this sample under the microscope revealed cancerous cells. This procedure, known as a biopsy, also showed that Craig's tumour was not of the 'small cell' type, a particularly malignant type that metastasized rapidly and was generally not treated by surgery. Frank Craig was fit enough for an operation, and soon after the diagnosis a surgeon removed a section of his lung. About a quarter of the lung cancer patients who undergo surgery survive this operation for five years, and about 15 percent for a further five years or longer. (At specialist chest units with experienced surgeons, this rate has changed very little since the 1950s). Unfortunately Frank Craig was one of the majority of patients whose cancers return within five years. In 1993 there were suspicious symptoms and in 1994 he was diagnosed with secondary growths. This is where the standard pathway ended. Other treatment modalities were applied: radiotherapy and some chemotherapy, and boundaries blurred between curative and palliative intentions. But Frank Craig's illness progressed and he became increasingly dependent, both on his wife and a growing number of helpers; he found it harder and harder to leave his bed. Frank Craig died two days after the fourth anniversary of the lung cancer operation, which, according to his wife, 'had at the time seemed so successful'.[7] Unlike Mary Benbow a century and a half before, he was killed not by the primary tumour but by secondary growths. As in Mary Benbow's case, this happened following a period of bodily disintegration.

Impotence and interventionism

This book is a history of lung cancer. While my case studies and examples are mostly from the UK, and to a lesser extent from the US, this is also a book about one of the great dilemmas of modern medicine: how do we deal with medicine's greatly improved powers to explain illness and to identify causes in circumstances where these improvements do not save patients' lives? The chances of a patient diagnosed with lung cancer in Britain in the early twenty-first century to survive the following five years are not a great deal better than they were 30 years ago. In fact, the standard treatment for non-small cell lung cancers has not changed greatly over the past half century. It is true, some types of cancer which used to be seen as hopeless a few decades ago – namely childhood cancers, lymphomas and leukaemias – have been viewed as curable since the 1970s, thanks to new and experimental regimes of chemotherapy.[8] Treatment outcomes for other types of malignant disease, such as breast or colon cancers, also appear to have improved significantly over recent years.[9] However, many other malignant diseases are much more like lung cancer, with only marginal gains in survival, seemingly defying the common assumption that money spent on cancer research is investment in survival time. The increase in life expectancy for an average American due to reduced cancer mortality between 1970 and 2000 was less than three months, in spite of the research efforts (and colossal expenses) associated with the 'War on Cancer'.[10] Translations from fundamental research into tangible improvements that changed treatment outcomes for the cancers with the highest incidence and death rates, have been rare. Recalcitrance has been the rule and not the exception.

Impotence in the face of illness has been a common experience throughout the history of medicine, especially when cancer was concerned. It was so before medicine became modern, and has remained so over the last 200 years. But even when there was good evidence that nothing very effective could be done, this rarely meant that nothing *was* done. Anne Davies, for example, a London widow and contemporary of Mary Benbow was bled and treated with leeches and belladonna plasters at University College Hospital until she 'sunk gradually' and died, with sounds emitted from inside her chest 'as if the whole lung were breaking up'.[11] Today, cancer treatment pathways still involve an arsenal of treatment modalities. This applies especially at the stages where, as in Frank Craig's case, routine treatments fail to deliver the desired responses – then highly experimental therapies may be tried, often with drastic side effects but limited benefits.[12]

As this book will show, the main routine treatment, surgery, has long been neither revolutionary nor new. It saves the lives of only a minority of lung cancer patients. Many patients are not suitable for an operation, because they are judged too ill to survive it, because their tumour sits in the wrong place or has already spread, or as in Frank Craig's case, because the cancer returns following the operation. Since the 1980s medical oncologists have promoted chemotherapy as an answer to the question of what doctors could offer to those who are unsuitable for surgery, or where tumours grow back. Desperate patients wanted something done, they argued, and were 'usually much more willing to undergo intensive treatments associated with substantial toxicity for what health professionals may see as minimal or no benefits in terms of the chance of a cure, prolongation of life or symptom relief'.[13] This book explores how doctors, patients, and others involved in frequently futile efforts to revolutionize the treatment of a recalcitrant disease, have dealt with this dilemma.

Lung cancer and recalcitrance

Lung cancer, the most common and, with 1.37 million deaths per year, the most deadly cancer worldwide, may be the most visible, but it is certainly not the only recalcitrant disease defying the ideology of progress and action that is so important for the image of modern biomedicine.[14] Survival rates of patients diagnosed with pancreatic cancer, for example, are terrible: three out of four die within a year of diagnosis. For liver cancer, not only are survival figures poor, there has also been a steady increase in incidence over the past three decades, associated by many with problematic drinking habits, although hepatitis infections are an equally likely cause.[15] With its particular history, however, lung cancer provides me with an especially suitable case study to illustrate how medicine over the past two centuries has come to deal with the phenomenon of recalcitrance. Let me explain why.

Lung cancer was largely invisible and considered rare until the mid twentieth century. How rare it really was is difficult to assess, as I discuss in Chapter 2. Because it could only be diagnosed reliably at autopsy and because phthisis, consumption, or tuberculosis were so dominant in the imaginations of both the wider public and those who noted the cause of death on the death certificates, a significant number of the deaths blamed on consumption may well have been deaths from lung cancer. In the first half of the twentieth century, with radical surgery increasingly established as a standard treatment for breast cancer and with surgery inside the chest being developed for the treatment of tuberculosis and other conditions of the lung, as I show in Chapter 3, the surgical treatment of lung cancer became a real possibility. By the 1940s, the available know-how and equipment in chest surgery and anaesthesia allowed surgeons to turn pneumonectomies and lobectomies (the removal of a whole or parts of a lung) into routine operations. This development coincided with epidemiological findings – followed by much public debate, which is the subject of Chapter 4 – demonstrating that, indeed, lung cancer incidence was increasing. That cigarette consumption was the cause of the increase was consensus by the early 1960s among experts outside what the British journalist Peter Taylor has termed the 'smoke ring' created and maintained by the tobacco industry.[16] A central question addressed in this book is what this discovery did to the treatment of lung cancer, a disease now increasingly firmly associated with a habit that had come to carry moral connotations – not quite as stigmatizing as sexual transmission, but also affecting the identities of sufferers.[17] Lung cancer was not just associated with other chronic lung conditions, it was increasingly linked with self-harm. As the eminent chest surgeon Thomas Holmes Sellors put it in 1966: 'Emphysematous and bronchitic elderly men who smoke heavily do not make ideal subjects.'[18]

The identity of a recalcitrant disease

As smoking became associated with delinquency and marginality, lung cancer did too. But in light of recent estimates that about 10 percent of lung cancers among men and 15 percent among women are not caused by smoking, this association was problematic. In absolute terms, there are approximately 9,000 men and 14,000 women per year in the US, for example, or 2,000 men and 3,000 women in the UK, whose lung cancer deaths are not related to smoking.[19] These are more deaths than are caused by leukaemia (23,500 per year in the US and 4,500 in the UK), which does not carry the stigma that lung cancer has acquired.

In the early 2000s, lung cancer research projects attracted less than 5 percent of UK cancer research funding, while accounting for 20 percent of cancer deaths; the figures for leukaemia were the reverse.[20] But contrary to the common assumption that lung cancer research has always been neglected, I will explore in Chapter 5 how the British Medical Research Council funded an important lung cancer treatment research programme from the 1950s.[21] Joint clinics involving chest physicians along with thoracic surgeons and radiotherapists then also promised to improve treatment outcomes. However, the disappointing results of the first clinical trials coincided with an increasingly negative outlook among surgeons treating the disease: they felt that no significant improvements of outcome could be expected, except through prevention; they were not likely ever to be able to save the lives of the majority of lung cancer sufferers, never mind how early the disease was diagnosed.[22] One of the few chest surgeons who maintained his optimism, J. R. Belcher characterized the attitude taking hold among his colleagues by the mid 1960s as 'almost unmitigated gloom'.[23]

Lung cancer appeared to defy all attempts to cure it with modalities that worked for other cancers. To be sure, radiotherapy had proven very effective as a palliative treatment, for example, reducing the immediate threat to life and limiting the terrifying swelling of a patient's face caused by a tumour obstructing the vena cava (the great vein returning blood to the heart). But attempts in the 1950s and 1960s to develop methods of radical radiotherapy for curing lung cancer yielded disappointing results. Chemotherapy, increasingly used since the 1970s where cancers recurred after surgery, exposed patients to the typical side effects but failed to keep them alive.[24] And surgery, while still the best possible option for a significant proportion of cases, more often than not failed to prolong patients' lives. While histology and tumour staging (used in increasingly standardized ways since the 1970s, as I discuss in Chapter 6) provided useful information, it was often unclear why a tumour did or did not reappear. The time of diagnosis did not always seem to make a difference, an observation that mocked the well established general assumption that if you caught it early, you'd cure it.[25] Thoracic surgeons frequently faced the spectre of hopelessness – in their practice as well as during professional meetings – and many of them looked to heart surgery rather than pulmonary surgery as a site of progress and source of professional pride and identity.[26] In the 1960s as today, few chest surgeons would describe themselves as lung cancer specialists, and this is partly due to what I call the recalcitrance of this disease.[27]

Writing history of medicine without progress

In the historiography of medicine, progress narratives easily outnumber histories of failure and disappointment; and I do not only refer here to histories written by doctors, or biographies of heroes of modern medicine, which often have hagiographic tendencies.[28] The oncologist Siddhartha Mukherjee, for example, introduces his remarkably successful 'biography' of cancer, *The Emperor of all Maladies*, with the story of a woman being treated for acute leukaemia, once an invariably deadly disease, which today is sometimes curable.[29] (While childhood leukaemia is the great success story of medical oncology, adult leukaemia is still quite recalcitrant.)[30] Mukherjee, however, like many other authors of popular books on cancer founds a progress narrative on the hope that successes in the treatment of a limited number of particular cancers can be extended to the great majority of malignant diseases. Adam Wishart's book *One in Three: A Son's Journey into the History and Science of Cancer* tells the story of illness and death of the author's father from metastatic prostate cancer, illustrating the impotence of modern medicine when faced with advanced malignancy.[31] But while this may be an example of recalcitrance, Wishart uses this story to introduce a narrative of progress in cancer research and therapy.

Perhaps surprisingly, historians of science and medicine have also rarely dealt with recalcitrance. Jason Szabo's insightful recent book on incurability in nineteenth-century France is a rare exception.[32] The subject has been addressed somewhat more extensively by sociologists of health and illness, especially those studying chronic illnesses and focusing on illness experiences rather than innovative medical practices.[33] But historians of science and medicine, by and large, tend to display a somewhat paradoxical attitude to progress. While our declared aim has long been to contextualize practitioners' claims about scientific and technological progress, we have usually chosen to focus on stories thought to epitomize such progress: innovative surgical procedures, artificial organs, the place of the laboratory in medicine, new medical technologies or other science-driven innovations such as medical genetics. Lung cancer is no exception: professional historians as well as the authors of insider histories or biographies have generally addressed this disease either in the context of what is often presented as a revolution in epidemiology that helped to establish the association with cigarettes, or by writing about innovative surgical procedures – the two success stories associated with this disease.[34]

Progress stories provide historians with a useful narrative framework, even where the main objective is to debunk these progress stories and point to the importance of context. But how does one write about recalcitrance? The history of lung cancer provides us with a good case to address this question; it illustrates how modern medicine deals with impotence and frustrations. In this book, laboratories will feature only on the margins. I will look at the attempts of clinicians to manage both the disease and the hopes of patients and wider publics, who expected breakthroughs and were more often than not disappointed. I will discuss surgery in contexts where it is not very glamorous. With a view to smoking, I will suggest that it was impossible to implement a satisfactory policy response to the epidemiological findings that identified smoking as the main cause of the worrying increase in lung cancer incidence in the 1950s.[35] What would have been an adequate response? Was a ban justified? Did prohibition work? Liberal societies usually turned to education as the best approach: if people knew what smoking did to them, surely they would quit; but it turned out that those who allegedly needed educating knew already. And many continued to smoke. Only slowly did the demographics of tobacco consumption shift, and those of lung cancer incidence followed after a decade or two.

Let me outline, very briefly, the structure of this book. Chapter 2, following on this introduction, deals with the emergence of lung cancer as a disease in the nineteenth century, explaining how what we now know as lung cancer turned from a form of 'consumption' into a specific disease entity. I argue that we cannot know how common lung cancer really was because only autopsies revealed if patients did not suffer from other forms of consumption. Chapter 3 continues to lay the foundations for the main argument of the book by sketching the history of the main treatment modality, pulmonary surgery, over the first half of the twentieth century. This is a fairly bloody story, focusing on technical innovations in anaesthesia and changes in the practice of chest surgery. In Chapter 4 I deal with an important turning point in the history of the disease: the debates, in the 1940s and 1950s, over the question whether there was in fact a lung cancer epidemic, and what caused it. I also discuss policy responses to the finding that, indeed, it was probably smoking that caused most lung cancers. Chapter 5 returns to treatment, in the 1950s and 1960s, introducing new approaches such as joint cancer clinics run by radiotherapists and surgeons, and the story of lung cancer clinical trials organized by the MRC unit that had also run the successful streptomycin trials,[36] ending with what Belcher characterized as unmitigated gloom among chest surgeons. In Chapter 6 I turn to the 1970s and 1980s, when expectations came to be more measured, in both meanings of the word. I also discuss the use of the new treatment modality, chemotherapy, along with attempts to develop an international staging system for lung cancer, as standardized and reliable way of distinguishing the majority of lung cancer patients who could not expect to survive the disease, even if operated, from the minority who were likely to benefit. I introduce the history of the unsuccessful attempts to establish efficient screening programmes for lung cancer and discuss why screening for lung and other cancers is so controversial. I also address the changing status of cigarette smoking. In Chapter 7, finally, I introduce the history of a charity, the first dedicated to this disease: the Roy Castle Foundation based in Liverpool which sought to overcome the hopelessness associated with lung cancer. I also discuss claims made by lung cancer advocates since the 1990s that lung cancer has become stigmatized, and underfunded as a consequence, and that this explains why the disease remains so deadly and recalcitrant to the present day.

2
Lung Cancer and Consumption in the Nineteenth Century: Bodies, Tissues, Cells and the Making of a Rare Disease

In 1912, the New York physician Isaac Adler, Professor Emeritus at the New York Polyclinic, Consulting Physician to the German, Beth-Israel, Har Moriah, and Peoples Hospitals, and to the Montefiore Home and Hospital, published what was probably the first monograph in English wholly dedicated to lung cancer. *Primary Malignant Growths of the Lungs and Bronchi* was based on a collection of individual cases described in the medical literature.[1] Adler extracted from this set of 374 cases the features that he believed to be characteristic of the disease, comparing them with experiences he accumulated in his own practice. The number may seem large for a compilation of case reports (the lengthy appendix to the book consists of a rather unwieldy set of tables listing every individual case), but it is small considering that this was the total number of reports Adler found in the medical literature published up to this point, not only in English but also in French and German, many of them as he put it, 'buried in dissertations and other out-of-the-way places'.[2] Roughly 15 years before the publication of Adler's book, Hans Pässler, a pathologist in the Silesian city of Breslau, had found only 70 cases that he thought could reliably be viewed as primary lung cancer.[3] With numbers so small, it was not surprising that many thought lung cancer was an extremely rare affliction.

Current approaches to lung cancer, still, usually start from the assumption that the disease was 'extremely rare' until the early twentieth century. But on what evidence is this statement based? It is almost trivial to say that lung cancer back then did not carry the same meanings as today (to start with, the association with smoking did not exist), but nevertheless, the assumption that this was one and the same

11

disease has frequently been made. This chapter has two main aims: first, I want to find out how lung cancer was diagnosed, discussed and made sense of in the nineteenth century. Second, I want to understand more generally what makes a disease rare, and if such retrospective statements, especially for lung cancer, are reliable. There will be no continuous narrative, as the story of lung cancer in the nineteenth century was a fragmented one, with developments – at different times – in Paris, Vienna, Berlin, London, and across the Atlantic in North America. For much of the nineteenth century, the story of lung cancer might be best characterized as a footnote in the histories of respiratory illness (especially tuberculosis) and of pathology. This does not mean that lung cancer as we define it today was not reasonably common. I will argue that we have no way of knowing the prevalence with any degree of certainty.

The first spotlight will be on new approaches in pathological anatomy and histology in the nineteenth century which, as Russell Maulitz put it, produced 'a roadmap of the human body decipherable by surgeon and physician alike'.[4] As I will show, the liaison of scholarly medicine and surgery was an important premise for understanding lung cancer as an object of nosology, as a diagnostic category.[5] Post-mortem examinations were fundamental to any claims about the rarity of this disease, and this is why I start this chapter in the early nineteenth century, when post-mortem practices first established in the Paris hospitals became more common outside France. In the eighteenth century, the symptoms of lung cancer would have occupied physicians, who most likely would have interpreted them as expressions of a 'fever' (then, of course, a respectable diagnostic entity). If categorized as a localized growth, lung cancer would have been a matter for surgeons (however, as I will show in Chapter 4, the surgical removal of tumours from inside the chest was a far way off and only became possible towards the mid twentieth century). The distinct disease identity of lung cancer that emerged in the early nineteenth century encompassed a set of symptoms – some local and others systemic (that is, affecting the body as a whole) – associated with a specific lesion: an anomalous growth in the lung, usually fatal, but not a tubercle.

I will look at three main developments in shaping understandings of the disease, informed respectively by pathological anatomy, cell biology and, towards the end of the nineteenth century, statistics. All three frames are central to Adler's book, and they remain important until the current day. Once the disease identity was established, with descriptions in textbooks and journals and an institutional basis in hospitals such as the Brompton Hospital for Consumptives and Diseases of the Chest in London or other teaching hospitals, its diagnosis became more frequent.

Patients, organs and statistics

In this and in subsequent chapters I will draw on accounts of individual cases and patient stories to illustrate how people made sense of the disease at the time. For the nineteenth century this is not difficult: we can rely on publications by doctors and surgeons, both books and articles in the new, modern medical journals. The genre of the case history was central to these publications over the course of the century, though.[6] But in the early nineteenth century they were mostly in the realm of natural history, describing and classifying the many kinds of afflictions which surgeons or physicians might be faced with in their careers (and listing therapeutic solutions). Later case histories turned more and more to analysis as pathological anatomy became widely established, dividing the patient's bodies into structural and functional elements and seeking to correlate signs and symptoms observed in the living patient with post-mortem findings.

Most of the patients we meet in nineteenth century British case histories were treated in London hospitals, some were in the larger provincial centres. Of course, there would be lung cancer sufferers outside the great cities, but they were note seen by doctors and surgeons who wrote down their stories in a teaching hospital or for a learned society (or both). London was the main centre of medical education in Britain, so most of the accounts came from the capital, fewer from Edinburgh and Glasgow, and fewer again from the provinces. These accounts were surprisingly rich, for the relevant biography of the patient was part of a proper diagnosis, along with observations made with new methods of physical examination, such as percussion and mediate auscultation. Most of the stories end with the deaths of the patients and with the pathologist's discussion of the lesions he found in their bodies.

Cancer of the lung was rarely suspected when the patient was admitted to hospital. That diagnosis was produced by new methods of diagnosis concerned with the location and function of the affected organs. The specifics of cancerous tissue in the lung became more important later in the century, informed by cellular pathology and new techniques and technologies to distinguish the histological origins of cancerous cells, and the relationships between normal and pathological cells and tissues. When interest focused on tissue samples and microscopy slides, the location of the cancer (in the chest or elsewhere) became secondary because the same types of malignant cells could appear in different organs. In the last part of the chapter I will look at statistics and the first attempts to account for an apparent increase in lung cancer incidence.

Pathological anatomy and growths in the lung

The first spotlight in this chapter is on René Théophile Hyacinthe Laennec's workplace in the Hôpital Necker, Paris. In Laennec's clinic in post-revolutionary Paris, post-mortem anatomy was combined with clinical observation and new diagnostic techniques. Here, lung cancer was given a name and its own place in a system that combined disease classification, diagnostics, and claims about its aetiology and nature. This is not the place for a detailed discussion of Laennec's work and of Paris Hospital Medicine.[7] I will only address those aspects that are important for the emergence of the new disease entity of lung cancer.

Laennec is best known for his invention of the stethoscope, the instrument that allowed physicians to use their ears to 'look' (as indicated by the term 'scope') into a patient's chest, and for laying the foundation of the modern ontological understanding of tuberculosis as a disease entity. For Laennec it was the specific form of consumption that is associated with tubercles in the lung (the term tuberculosis is said to have been coined by the German physician Johann Lucas Schönlein with reference to these lesions). It is not unusual in the history of medicine that new understandings of disease resulted from technical innovations and innovative practices, and I will deal with several occasions in this book where this was the case for understandings of lung cancer.[8] Laennec's classificatory system built on the new pathology of tissues introduced by M. F. Xavier Bichat. The new understanding of consumption that emerged from Laennec's clinic and the circles of colleagues and friends he interacted with, relied on a combination of careful clinical observation (supported by Laennec's new instrument, the stethoscope) with routine post-mortem examinations.[9] What defined the disease in the classificatory system promoted by Laennec and his followers was no longer a (potentially infinite) set of symptoms that the physician considered in the context of the patient's biography, but the existence of a finite number of specific disease markers in the body, found after death and assumed to be present already in the living, causing the illness. As far as chest diseases were

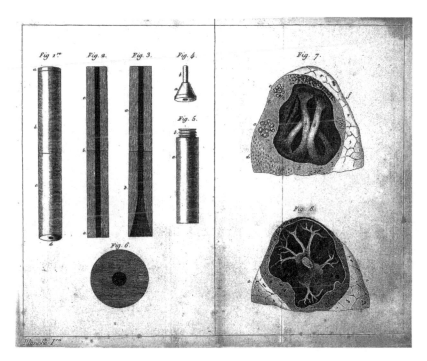

Figure 2.1   Stethoscope and Lungs. Illustration from Laennec's *De l'auscultation mediate*. Reproduced with permission.

concerned, the new definition also relied on meticulous differentiation between lesions associated with tuberculosis and other lesions that may be associated with related symptoms and picked up by the same diagnostic methods. Tubercles were one form of new growth in the lungs, but there were other growths, for example the 'cerebriform' or 'medullary' tumours, or encephaloids, all named so because of their apparent similarity to brain tissues.

Laennec first described the encephaloids in an article in the *Dictionaire des Sciences Médicales* in 1815.[10] His friend and collaborator Gaspard-Laurent Bayle had observed this form of tumour in a post-mortem examination a few years earlier and identified it as scirrhus or cancer.[11] Bayle classified it as a variety of consumption and named it 'cancerous phthisis'. Laennec disagreed with Bayle, classifying this type of tumour as a form of 'accidental production' and naming it after what he thought was a striking resemblance to brain tissue. 'Accidental production', according to Jacalyn Duffin, was 'the jewel in the crown' of Laennec's classification of organic lesions, which he first published in 1805 and then with little modification in 1812 and 1822, and which was fundamental to his conception of disease.[12] He used the term to describe tissues that seemed to have formed without obvious cause; they could be 'analogous' productions (composed of tissues found in the healthy body) or 'non-analogous' productions. The 'accidental productions' corresponded loosely to what medical science later came to term 'neoplasm', with its subdivision into 'benign' and 'malignant'. Crucially, the non-analogous accidental productions included the 'tubercle' along with growths that we today would identify as cancers, the hard *squirrhe* (in English usually spelled scirrhus or schirrhus), the soft *encephaloïde* (encephaloid), and the *melanose* (melanoma).

Some of the claims and concepts that Laennec promoted along with his classificatory system were quickly forgotten. Others were never completely accepted. The great Vienna pathologist Carl von Rokitansky, for example, never accepted Laennec's premise that cancer had local origins and was convinced that tuberculosis and cancer would never occur in the same individual; for Rokitansky they were both systemic diseases but triggered by fundamentally different 'dispositions'.[13] Laennec's terminology, however, was used in case reports on cancerous growths found in patients' lungs until well into the age of cellular pathology. For the duration of the nineteenth century lung cancer was an (often unexpected) autopsy finding, a chest disease that was neither phthisis nor one of the other more common afflictions of the chest, pneumonia, bronchitis or emphysema.

The diagnosis of encephaloid growths and pathological anatomy in Britain

An English translation of Laennec's *traité de l'auscultation mediate* by John Forbes was published in 1821.[14] Forbes, a Scotsman by origin, was Physician to the Penzance Dispensary, Secretary of the Royal Geographical Society of Cornwall, and belonged to a small circle of reform-minded doctors who promoted training in pathological anatomy as a means of turning the British surgeon-apothecaries into scientifically informed general practitioners. Forbes had not himself studied in France, but had heard about Laennec from James Clark, a childhood friend and also a reform-minded physician. Forbes had served as a naval surgeon from 1807, then studied medicine at Edinburgh and graduated in 1817. Later he held appointments first as a dispensary and then an infirmary physician in Chichester. He also launched and edited two journals, in 1832 the monthly *Cyclopaedia of Practical Medicine*, which according to R. A. L. Agnew provided a forum for the best medical writers in the British Isles, and following the sale of the *Cyclopeadia* in 1836, the *British and Foreign Medical Review*, subtitled *A Quarterly Journal of Practical Medicine*. In 1840 he moved to London, where in 1846 he was appointed as one of two consulting physicians to the Brompton Hospital for Consumption and Diseases of the Chest.[15] Forbes abridged Laennec's book and changed its structure, separating pathological anatomy from diagnosis and omitting most of the physiological content, which had been most important to Laennec, along with much of the detail in the case studies, which Forbes deemed irrelevant to his projected British audience.[16]

Forbes' translation was published just as the British medical world was becoming aware of the flow of English medical students to Paris, to experience the new anatomy-based, practical teaching methods and the easy access to dead bodies for autopsies.[17] The medical schools at the Scottish universities embraced the new pathological anatomy fairly quickly. In England, however, with its conservative Royal Colleges and proprietary medical schools, the reception was slower. Gradually, however, the links between pathology and clinic grew closer, with Laennec's techniques of physical diagnosis providing a crystallization point. Cadavers for training purposes remained a rare commodity for some years to come, but post-mortem examinations of patients treated in British teaching hospitals (and not only there) became more common, supported by the growing number of pathological societies and of new medical journals and gazettes that encouraged publications on interesting cases.[18] Probably the best-known of these new journals was the radical *Lancet*, launched in 1823 and promoting medical reform; another example was Forbes' *Review*.

Post-mortem examinations were rare in the English-speaking world before this period, and the uptake of the new practices associated with pathological anatomy differed between hospitals. The *Lancet* specifically attacked the Middlesex and St George's Hospitals for rejecting post-mortems. The author of a polemical article in 1825 quotes a surgeon at the Middlesex: 'There is no use in pulling dead bodies about'.[19] Dr Pelham Warren, a physician at St George's, was quoted as complaining that 'this d-d morbid anatomy will spoil the practice of physic'.[20] In other hospitals, such as Guy's, St Thomas's and St Bartholomew's, post-mortems were more common. And they were becoming ever more common as the principles of pathological anatomy were increasingly firmly established in British medicine.

Forbes' translation of Laennec's *Traite* was not the only book on the new techniques of physical diagnosis published in English. Forbes himself published a small treatise on stethoscopy in 1824, with English cases.[21] Many others followed, some on pathology and morbid anatomy, others specifically on auscultation or percussion.[22] Some of the observations of encephaloid tumours in the lung in Britain were made while authors were preparing books on auscultation and the new techniques of physical diagnosis, others in books dedicated to the diagnosis and treatment of diseases of the chest. One example of the first kind is the case of Mary Benbow, with which this book opens. An example of the second kind is the chapter on lung cancer in a book by William Stokes, which I will be looking at first, as it was published a few years earlier and allows more general observations on the diagnosis of this disease at the time.[23] Stokes's account illustrates the difficulty of diagnosing lung cancer and shows that the disease was important primarily because it was not tubercle, pneumonia or empyema. His case descriptions are unusual in as they do not provide us with any details on the patient's stories and focus entirely on clinical and post-mortem observations. All case descriptions were attempts to let readers take part, in a virtual way, in the process of clinical observation, decision making, and attempts to reconcile clinical observations with post-mortem findings. And all the cases of lung cancer show that the patients would have been diagnosed as consumptives had their bodies not been subjected to post-mortem examinations.

William Stokes, or, how to diagnose lung cancer rather than tubercle, pneumonia or empyema

Stokes describes a number of cases of lung cancer in his 1837 book, which he thought of as representative of two forms of the disease:

> in the first a degeneration of the lung occurs, and the organ is transformed into a cancerous mass without the production of any tumour. In the second, the schirrous or encephaloid matter forms a tumour, at first external to, and ultimately displacing the lung. In neither case can we apply any direct diagnosis; and I do not know how the first could be determined with certainty. The symptoms are always obscure.[24]

The first case, a man of 36 years of age showed the typical signs of what later became known as superior vena cava obstruction (the blockage of one of the great veins in the chest cavity), causing the swelling of smaller veins on the surface of his upper body and a bloated appearance of the neck and face – a complication that is typical for lung cancer. He suffered from cough (sometimes coughing up blood), hoarseness and dyspnoea (breathing difficulties). Physical examination revealed that 'the whole side sounded dull, yet without the accompanying physical signs of a great empyema on the one hand, or of pneumonia or tubercular solidity on the other'.[25] Stokes's colleague, Dr Graves, who also treated the man, observed three tumours on his body, immediately under the skin, which increased in size 'with great rapidity'.[26] The man died and the dissection revealed, in place of the right lung, 'a solid mass, weighing more than six pounds'.[27] The mesenteric glands had also been replaced, it seems, by a tumour that consisted of 'the same cerebriform substance as that observed in the chest'.[28] Stokes concludes: 'there was here a group of phenomena irreconcilable with those of pneumonia, phthisis, or empyema, but which were explained by the condition of the lung'.[29]

The second case was that of a woman, aged 30, who also had difficulty breathing and suffered from cough and mucous expectoration. Furthermore, she complained that she found swallowing difficult, feeling as though she had a lump deep down in her throat. Based on a physical examination involving auscultation and percussion (a diagnostic technique that had the physician 'drum' on the patient's chest with the tip of his fingers, listening to the sounds this produced), Stokes concluded that 'the superior portion of the left lung was solid, in all probability from tubercle'.[30] The patient died in the night following her admission. The dissection of her body revealed that 'the left lung was found compressed ... by an extensive encephaloid tumour' that enclosed the trachea and the oesophagus and that 'answered exactly to Laennec's description of the non-encysted cerebriform masses'.[31]

The diagnosis in a third case, of a man of 45 years of age, was also full of difficulties. Again there were cough, breathing difficulties and unusual sounds. Some observations pointed to tubercle, others to an aneurysm (a widening of an artery looking a bit like a balloon), others were irreconcilable with either explanation. When the patient died and the body was dissected, Stokes found a large, globular tumour in his left lung, which surrounded and compressed a portion of the pulmonary artery.

In concluding the chapter, Stokes suggests that cancer of the lung might be suspected in patients that show 'evidences of simple solidification without the signs of pneumonia or tubercle' or 'where there were evidences of an intra-thoracic tumour'.[32] The term tumour is used here only to refer to a clearly defined mass, as in the third case, which did not take over the whole lung. Stokes also compares his cases with others reported by Andral and Bayle in Paris, and by Robert Carswell at the medical school of the new University of London.[33] There is no discussion in Stokes' book of the nature or origin of the cancerous tissue. His interest was diagnosis.

Differential diagnosis in Victorian London

I have presented the case of Mary Benbow, written up by Dr Henry Marshall Hughes of Guy's Hospital, London, on the opening pages of this book. Another case was that of Anne Davies, a patient of Dr John Taylor's in the most progressive of the capital's teaching hospitals at University College.[34] The stories of Mary Benbow and Anne Davies provide us with more detail on the patients than the cases reported on by Stokes. Anne Davies, we learn, kept a milliner's shop, was 41 years old and had been a widow for ten years, when she was admitted on 12 October 1841. Taylor found Mrs Davies to be of 'sanguine nervous temperament'; she had dark hair and eyes, and her habits were regular. A visitor told the nurse that she had been imprisoned lately for debt and suffered from anxiety. One of her sisters was said to have cancer of the womb and was living. That these details are mentioned points to the roles assigned to psychological factors and heredity in the aetiology of the disease. Mrs Davies was apparently in good health until the previous February when she moved into a house in Islington which turned out to be damp, and she attributed her illness to this fact. Four months before her admission to the hospital she had suddenly felt a severe pain in the lower back, which was so acute that she fell down. The pain had not gone away and was sometimes stronger and sometimes less bad. Then, a month later, she suffered what Taylor explained as an attack of inflammation of the right side of the chest, with pain, breathing difficulties, and a violent, occasionally bloody cough. Like Mary Benbow, she was bled to alleviate these symptoms. When the cough continued and the pain in the back got worse, making it difficult for her to walk, she sought admission to the hospital.

On admission she was found to be generally emaciated, anxious, her complexion sallow and with a pale-yellow tinge (symptoms that Taylor retrospectively thought to be typical of cancer). Her appetite was impaired and she was unable to eat any 'animal food' without throwing up.[35] Her pulse was frequent and rather small and soft. There was no swelling in the lower back, where the pain was most severe, but a tumour could be felt in the abdomen, which was apparently fixed to the vertebral column. Taylor examined this carefully, noting details about pulsation and the sounds of blood flow, presumably with the help of a stethoscope. Mrs Davies also complained of a sense of obstruction in the gullet and had difficulties swallowing. But there were no cough (with or without blood), no breathing difficulties, and no pain in the chest. He observed the rule that he always followed, as he claimed, of examining every organ in every case, not only those where he expected to find disease.[36] The right side of her chest seemed flatter than the left, but the difference was not measurable. The movement of the right chest was less marked than of the other side, and on percussion the sound was dull. On auscultation, the natural respiratory murmur was replaced by the sound of 'bronchial respiration'. In the lowest parts of the right chest it was hard to hear anything. It was clear that the right lung was diseased:

> The nature of the disease upon which this increased density depended, was, however, still to be determined. ... There were four diseases, amongst which, chiefly, he had to seek for an explanation of the existing symptoms; namely, pleurisy, pneumonia, phthisis, and some morbid growth, such as cancer of the lung or adjacent structures.[37]

Taylor described the sounds he heard in the patient's chest in great detail. He also noted a lot of details which do not seem immediately relevant to us and to the complaints that made Mrs Davies seek admission to the hospitals, such as bowels that are 'rather confined' and that she had 'generally some pain in the head'.[38] The aim here, it seems, was to record a set of observations that was as broad and general as possible, and that illustrated the way the clinician reasoned, inviting the readers to come to the same conclusions.

Over the following six weeks the condition of Mrs Davies got gradually worse, and there was not much the doctors could do about it. The pain in her back got stronger and more persistent, and she now also felt it in her legs, she became even more emaciated, and the breathing difficulties came back and caused her much anxiety. She was treated with leeches and belladonna plasters to relieve the pain, and with hot compresses to the spine. A prescribed half pint of ale made her rather delirious and had to be omitted. Breathing became more and more difficult, and a gurgling sound could be heard now on both sides of the chest. Dr Hare, 'a very intelligent pupil' of Dr Taylor described a sound 'as if the whole lung were breaking up'.[39] On 27 November she 'sunk gradually' and died in the late afternoon.

Forty-one hours after her death, Dr Taylor and his colleagues 'inspected' her body. They found that the right chest contained about eight ounces of a thickish, dirty grey fluid. The lung looked abnormal and smelled offensive, and 'a foreign deposit was copiously diffused throughout the substance of nearly the whole lung'.[40] About the root of the lung was what Taylor described as 'a considerable quantity of encephaloid tissue'. They also found 'several cancerous masses' in the kidneys and a tumour of about the size of a large orange at the lower end of the vertebral column, which 'presented the characters of encephaloid disease, and on pressing it a creamy liquid exuded from numerous points'.[41] The 'foreign matter diffused through the right lung' was examined under the microscope by Taylor's colleague, Dr Walter Hayle Walshe, Professor of Pathological Anatomy at University College, Physician to University College Hospital and to the Hospital for Consumptives and Diseases of the Chest, who at the time must have been working on his book on *The Nature and Treatment of Cancer*.[42] This foreign matter was 'found to present the microscopical characters of cancer'.[43] Importantly, however, microscopical examination is rather marginal to Walshe's book, and what there is, is very different from the classifications we find in twentieth century textbooks.

Cellular pathology: A revolution?

The examples above will suffice to illustrate what lung cancer was until well beyond the mid nineteenth century, a disease of the lung that was neither tuberculosis, nor pneumonia, nor empyema, for which there was no cure, and that with certainty could only be diagnosed after death, when usually the cancer had taken over the whole lung and was often also found in other places. This led to discussions about the primary site, and some authors argued that no cancer had its origins in the lung; all lung tumours were secondaries to cancers growing elsewhere in the body. The examples also illustrate how the literature on lung cancer was one of identifiable, individual cases. There were, however, other developments, namely the use of the microscope to study cancerous tissues and developments in statistics, to which I will turn now, concentrating on those aspects that are important to understand the history of lung cancer.[44]

The cellular concept of cancer was exciting because it provided the foundations for new explanatory concepts as well as new diagnostic and therapeutic approaches. The cell theory had its origins in microscopical observations made by Theodor Schwann in Berlin in the 1830s, while working with the influential physiologist Johannes Müller.[45] Schwann used a new, much improved compound microscope with an achromatic lens, and the interpretation of what he saw was informed by recent work by the botanist Matthias Schleiden.[46] In 1838 Johannes Müller himself published a book *Ueber den feineren Bau und die Formen der krankhaften Geschwülste* (On the fine Structure and Morphology of Malignant Tumours). His pupils Robert Remak and Rudolf Virchow refined and expanded the Schwann-Schleiden cell theory into a complete set of pathological practices and concepts.[47] In addition to the new microscopes, new techniques for staining, fixing and slicing tissues into ever thinner sections played an important role in this development. Virchow published his book on *Die Cellularpathologie* in 1858 and a compilation of 30 lectures in which he applied the new concepts to cancer between 1863 and 1867.[48] 'Cellular pathology' was based on the assumption that all cells were formed from parent cells, and he aimed to explain all disease in terms of cellular change, 'with unequalled success to oncology', as the historian of medicine Erwin Ackerknecht argues.[49] Most modern tumour definitions are based on cellular pathology and much of the terminology in oncology up to the present day refers to the assumed cellular origin of cancer cells rather than the tumour morphology which had interested Laennec. But how did the transition from Laennec to Virchow, as it were, occur in practice, and what were the implications specifically for lung cancer?

John G. Gruhn, in a chapter on the history of the histopathology of lung cancer, along with many other modern writers on the history of cancer, assumes that with his influential 1858 text on cellular pathology Virchow 'swept away the humours and dyscrasias that had obscured progress for centuries'.[50] In fact, the transition was far less clear, and we probably should not expect practices to change from one year to the other in response to a book published by a Berlin pathologist. In the eyes of many doctors, Laennec had already separated pathology from the traditional system of the four humours. But some influential pathologists, like Rokitanski in Vienna, found the practices of pathological anatomy not at all incommensurable with theories of

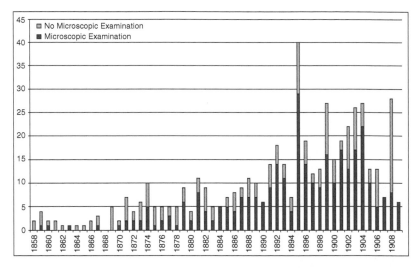

Figure 2.2 Lung cancer cases listed in Adler's book with information on whether or not a microscopic examination was performed. Only cases from 1858 onwards were included, and only those where Adler lists the publication date.[51]

humoral imbalances (dyscrasias) as causes of disease.[52] And with early cell theory assuming that cells formed from the surrounding fluid (the blastema), why should it not be possible to combine an interest in microscopy and cells with humouralism? In any case microscopic observations were not undertaken routinely after every autopsy for many years. Figure 2.2 shows this for the lung cancer cases Adler reviewed for his book.

In practice, as we can see here, the old was not simply swept away by the new. Rather, microscopy and cellular pathology added new interpretative devices to the toolkits of physicians, surgeons and pathologists, but they also continued to use the old ones.[53] Why should they not? Case histories, clinical judgement and diagnostic decisions based on gross morphology and the appearance of lesions continued to be central to diagnostic practice well into the age of cellular pathology. Detailed, individual case histories continued to feature in journal articles on lung and other cancers, while microscopic observations, if at all, were covered in a sentence or two. Even where these were undertaken, the language and terminology used was sometimes Laennec's, with carcinomas characterized as encephaloid. The decision between sarcoma and carcinoma was often made based on the gross appearance of a tumour rather than microscopic examination. The histological distinction that is most important in the diagnosis of lung cancer today, that between oat cell (or small cell) and non-small carcinoma only came into routine use in the mid twentieth century. Until the 1920s, oat cell carcinomas were widely assumed to be lymphosarcomas. There was no standardized and generally agreed terminology when Adler published his book in 1912. He described this as 'almost intolerable confusion in the nomenclature'.[54]

Take the case of a young sailor, for example, J. H., aged 25, admitted to the Seamen's Hospital on board the ship *Dreadnought* on 14 September 1866.[55] Dr Ward, apparently one of the attending physicians, found this case worth reporting as the 'diagnosis of cancer is often attended with no small difficulty, especially as regards its distinction from tubercle' and 'well-recorded instances are always valuable'.[56] H. had been feeling ill for 15 weeks and attributed this illness to hard training for a rowing match. First he had felt 'merely uneasiness about the chest and shortness of breath', but later he began to cough up blood. The physical examination on admission revealed the symptoms we already encountered in other cases: unusual sounds in some places of the chest and the absence of respiratory murmurs in others. H. remained in hospital until mid October, by which time he also suffered occasional fainting fits. On 19 October he left for lodgings in Greenwich, visited from time to time by the resident medical officer of the *Dreadnought*, Mr Leach. H. grew increasingly weaker, suffering distress and pain in the left side of the chest and great dyspnoea (shortness of breath), and died on 26 November. Mr Leach performed an autopsy 48 hours after H.'s death, finding two-thirds of the left lung 'involved in, and incorporated with' a large, medullary tumour. Only the fluid oozing from this tumour was examined under the microscope, revealing 'numerous small nucleated cells'. There is no further discussion of this finding in the report in the *Lancet*. In other cases, for instance that of the 40-year old labourer Patrick K., who was treated by Dr Charteris at the Glasgow Royal Infirmary in 1878 and whose 'symptoms during life were very obscure', there is no mention of any microscopic examination at all.[57]

There was also no clear transition in the terminology that clinicians used in their case reports. The compilation of case reports in Adler's book includes many fungiform and encephaloid tumours after 1858. In many cases old and new terminologies and concepts were used in combination. An 1887 article in the *Transactions of the Pathological Society of*

London talks about a bronchial growth 'found to be a carcinoma con-sisting of large cells with large, deeply-stained nuclei embedded in a fibrous stroma forming a hard mass, so that it should be considered a glandular scirrhus'.[58] Microscopic examination of samples secured during autopsies and observations on the cellular nature of particular tumours, it seems, were simply added to the set of practices that were already in place, and not consistently on all occasions. However, this did not mean that the older practices were abandoned.

Diagnosis in the age of cellular pathology

There was a fundamental difference between the breast cancer cases, which accounted for many of the samples seen by pathologists, and lung cancer. Breast cancer was quickly getting established as a target for surgery, while lung cancer was not, remaining in the domain of physicians, some of them specializing in chest problems. Breast tumours were located near the surface of the body, and often they were ulcerating, breaking through the skin by the time patients went to see a doctor. Lung tumours, in contrast, were inside the rib cage and not accessible for routine surgical procedures until well into the twentieth century, as I discuss in the next chapter, when new technologies in anaesthesia and infection control were available, which made such interventions feasible and paved the way for the establishment of the specialty of thoracic surgery. In most lung cancer cases, as we have seen, a tumour was only identified as a cause of chest problems after the patient had died, and this did not change for decades. Not surpris-ingly, this had consequences for the ways in which samples for exam-ination by pathologists were collected. Gabriel Andral, a rival of Laennec, had suggested that the examination of samples of sputum (the matter that patients coughed up), looking for encephaloid or cirrhus tissue, might provide a way of diagnosing lung cancers as early as 1821.[59] But even with the new microscopes, sputum examination never turned into a reliable and routine method of diagnosing lung cancer. While there were occasional reports in the international medical press in the closing decades of the nineteenth century on lung cancer cases diagnosed by way of sputum examination, a book on sputum microscopy published by G. H. Mackenzie in Edinburgh in 1886 did not even mention lung cancer.[60]

Occasionally lung cancers were treated by surgeons. The following case, however, suggests that this was unusual. Surgeon-Major Curran of Warrington reported this case in the *Lancet* in 1880, admitting to have completely misdiagnosed a cancer of the lung.[61] A ten-year old boy had developed a painful swelling on the outside of his chest after being hit by the shaft of a grocer's truck on 24 April 1879. Curran, assuming that he was dealing with a local abscess of some description, punctured the 'tumour' a number of times, every time drawing significant amounts of blood. The boy, however, seemed to get more and more poorly. 'Fancying that something more might be done for him, or rather reproaching myself for not having interfered more actively before', Curran 'asked two friends from the town to see him with me on the 2nd of August, and they, after carefully measuring the growth and examining the lad himself, advised me to leave it and him alone.' Eventually the boy also exhibited swollen veins in the feet and upper body and the tumour started to give out 'a sickly cadaveric odour'. Curran consulted with another doctor friend, 'who kindly came some distance to see him with me'. Curran's report is not that of a cold, detached observer. It is obvious that the suffering of the boy, who grew increasingly exhausted, troubled him. But his patient

> would not take any medicine, and as to subcutaneous injections of morphia, or other local application, it was quite out of the question. He cried out as I approached him on the morning of the 6th [of September] that he was dying, and the general appearance did not belie this assumption.[62]

On the next morning the boy lost about a pound and a half of bloody liquid from the tumour, and this led to a crisis, his breathing becoming gasping and his pulse turning feeble. 'Fortunately', according to Curran, 'the sensibility had been so numbed by the bleeding as to minimise the suffering'. He finally died, 141 days after the injury that was blamed for his illness.[63]

The autopsy, however, revealed that it was not the accident that had slowly killed the boy. The surgeon found in the boy's chest 'a malignant mass of medullary cancer', weighing 4 pounds and 14 ounces, 'which had eroded the bodies of the seventh, eighth and ninth ribs, and escaped, by the entire destruction of the eighth, through the thoracic wall.' Curran sent a section of the tumour to a colleague, Dr Klein for examination under the microscope, quoting six lines of rather technical details from Dr Klein's findings in his report without further commenting on these. In his concluding remarks Curran admits to having completely misjudged the case:

> Acknowledging with some feeling of compunction, but without any shame, my own gross, palpable, and profound misconception of the 'point' of this case *ab initio*, ay, and almost *usque ad finem*, I must observe that I did not stand alone in this respect, and that this was only the third example of its kind that has come under my notice during a professional career that now exceeds twenty-six years.[64]

But not only lung cancer was a rare occurrence: 'the other instances of cancer that I have seen ... were two of the liver, one of the parotid, and one of the testicle'. He and his colleagues were misled, he argued, by the original blow and the bleeding, and they 'never even suspected that the external tumour was a continuation of the lung'. Other typical symptoms associated with malignancies, such as the so-called 'cancerous cachexia' or 'that peculiar sallow complexion' were 'simply conspicuous by their absence'.[65]

These cases show that cellular pathology initially did not make much of a difference, as far as the diagnosis of lung cancer was concerned. And even if Curran had recognized that what he thought to be an abscess was in fact the extension of a tumour inside the rib cage, there was not much he could have done about this. In the absence of an established classificatory scheme for such tumours, cellular pathology did not make much difference even for pathologists undertaking post-mortem examinations. While the techniques of physical diagnosis, as we have seen, did not allow the reliable detection of lung cancer, over the following decades a number of promising diagnostic techniques were developed, including bronchoscopy and X-ray diagnosis. But these were not used routinely before the 1930s, not even in specialist hospitals.[66] I will discuss these techniques and the growing importance of specialist chest hospitals as sites of innovation in Chapter 3. Meanwhile, Surgeon-Major Curran's lack of experience with all forms of cancer brings me to the final section of this chapter and the question of statistics: how rare or how common was lung cancer in the nineteenth century? Was it increasing in frequency? Was cancer in general getting more common, as some suggested, due to the strains of modern life? And how reliable were the figures, the statistical methods, on which such statements were based?

Counting cancers

Some of the analytical tools of modern medical statistics can be traced back to roughly the same time as the new microscopic and histological methods that provided the foundation for cellular pathology, the mid nineteenth century. As was the case for pathological anatomy, the roots lay in post-revolutionary Paris, and knowledge and approaches travelled from there to Britain and elsewhere, where they were appropriated and adapted for local purposes. Historians of medical statistics often name Pierre Charles-Alexandre Louis with his 'méthode numerique' as the first modern epidemiologist.[67] Like Laennec, Louis was a well-known teacher and attracted many students from overseas who introduced his methods, for example, to Britain, where they met with a Victorian society fascinated by classification and statistics. One of Louis's British students was William Farr, who was appointed in 1839 as compiler of statistical abstracts to the General Register Office (GRO), an agency for central data collection that had been established in the previous year.[68] Cancer, to be sure, was not among Farr's main concerns. Like most 'sanitarians' of the mid nineteenth century, he was more concerned with highly visible disease events such as cholera epidemics or the big killer, tuberculosis. During his tenure at the GRO, Farr collaborated with other pioneers of medical statistics, such as John Snow or Florence Nightingale. He devised methods of data collection, introduced categories of occupation and of disease, and published reports. He structured the data collected by the GRO by grouping people by age and by occupations, which he categorized into five classes and 18 orders. He also developed a taxonomy of disease, which he continued to refine over several decades, grouping illnesses into five classes: zymotic, constitutional, local, developmental and violent. Cancer in this system was classified as a constitutional disease, along with gout, dropsy (a pre-localist term for what today would be diagnosed as kidney disease[69]) and tuberculosis. The fifth class, 'violent', included accidents, battle deaths, homicides, suicides and executions, indicating that this was in fact a taxonomy of causes of death rather than disease. This, again, reflects the central role of the post-mortem in nineteenth century medicine.

Towards the end of the century, epidemiologists increasingly turned their attention to cancer. In 1893, the Honorary Secretary of the Institute of Actuaries, George King, and the then Medical Officer of Health for Brighton, Arthur Newsholme, published a paper in the *Proceedings of the Royal Society of London*, 'On the alleged Increase of

Cancer'.[70] Newsholme, like Farr, was a pioneer of modern epidemiology, and Isaac Adler cites him in his book as one of his main sources
of evidence for his, Adler's reflections on possible increases in the
incidence of cancer in general and lung cancer in particular.[71]
'During the last few years', King and Newsholme began, 'the minds of
medical men and of the general public have been exercised over the
rapid and striking increase in the mortality from cancer, as shown by
the statistics contained in the Registrar-General's Annual Report'.[72]
The registered death rate from cancer in 1891 was 2.7 times as high
as the average for the decade 1851 to 1860 for men and 2.2 times as
high for women. 'That cancer has really increased in this country',
they continued, 'appears to be now generally assumed in medical
circles'.[73] 'Medical opinion', Newsholme suggested in another paper
six years later, in 1899, 'has been exercised with the notions that the
increase is due to the increased strain of modern life, or, in the alternative, that it is simply due to the larger number of persons who in
these days of improved expectation of life survive to the "cancer
ages"'.[74]

Statisticians and epidemiologists like Farr and Newsholme worked
with data that came from two main sources: the census, taken every
ten years, and the civil registration of births, marriages and deaths,
sometimes supplemented with life insurance data. The census data had
been fairly inconsistent until the task was transferred to the GRO in
1841, when the census was redesigned to improve its scientific utility.
The system of civic registration was created in 1836, requiring the registration of all deaths in England and Wales with a local Registrar. The
registration system, like the early census enumerations, was based on
the geographical units of the Poor Law. From the beginning, the cause
of death was required at the time of registration, but initially anyone,
whether medical practitioner, coroner or layperson, was entitled to
decide what caused a death, which rendered the data gained somewhat
incoherent and limited its use for systematic statistical analysis. Farr
and the Registrar General campaigned for a certification of deaths by
medical practitioners, fairly successfully, as by 1872, 92 percent of the
deaths in England and Wales were so certified.[75] The 1874 Births and
Deaths Registration Act made the medical certification of deaths com-
pulsory. Nevertheless, statisticians continued to deplore the sometimes
poor quality of the data so gained. Also, this system was not ideal for
statements about illnesses that did not end in a patient's death, and
this did not escape the attention of contemporaries. As early as the
1840s there were demands for the registration of sickness as well as
deaths, in order to allow statements about morbidity as well as mortal-

ity. With the 1889 Infectious Disease (Notification) Act this became reality, but only for diseases classified as infectious.

But back to cancer: had its incidence really increased since the mid nineteenth century? King and Newsholme came to the conclusion that it had not, and that the apparent increase was simply due to improved diagnosis and registration, especially of cancers that, like lung cancer, affected internal organs. The data from England and Wales alone was insufficient to support their hypothesis, as death certificates did not include information on the organs affected by cancer. Such data, however, was available for the German city of Frankfurt on Main, and this data showed that the increase was exclusively in cancers that they classified as inaccessible (which included the respiratory organs). As doctors learned more about these and as methods of diagnosis improved, King and Newsholme argued, these cancers were more frequently identified as causes of death. This explained why the increase was more pronounced for men: the main female cancers (in breast, uterus or vagina) were all classified as accessible and were frequently detected. King and Newsholme did not find it plausible that only inaccessible cancers were on the rise while the incidence of the more accessible ones remained the same. Newsholme's arguments did not convince everybody, however. The surgeon W. Roger Williams in his 1908 *Natural History of Cancer*, for example, suggested that 'taken in its entirety, the increase is so enormous as to make this explanation quite far fetched'.[76] Williams argued that the increase in cancer was linked to a decrease in tuberculosis incidence, as poverty, overcrowding and bad sanitation, which led to tuberculosis, did not favour the genesis of cancer, 'which flourishes most under just the opposite conditions'.[77]

According to the Frankfurt statistics, praised as exemplary by King and Newsholme, only about 1 percent of all cancer deaths in the city in the 1880s (1881–1889) were diagnosed as due to cancers of the respiratory organs (14 deaths out of 1,260), 2 percent of male cancer deaths (nine out of 454) and a bit more than 0.5 percent of the female deaths (five out of 806).[78] However, there were 77 male and 72 female deaths from cancer at non-specified sites, and it is possible that some of these also had their origins in the lungs. Still, it is clear that the diagnosis of cancer of the lung was rather rare. It is impossible, however, to make statements about the actual incidence. We simply cannot know.

Conclusion: A rare disease?

To conclude this chapter, let us return to Isaac Adler and his 1912 book on lung cancer. Adler followed Newsholme's line of argument, suggesting that lung cancer was probably not as rare as previously assumed, but, rather, that it was under-diagnosed, and that therefore any apparent increase in recent years had to be attributed to doctors becoming more aware of the existence of this disease. Still, for most of his colleagues, he suggested, 'the ubiquitous tuberculosis ... is ever ready to furnish ... a comfortable and satisfactory diagnosis' for any affliction of the respiratory tract.[79] But who could blame them, as 'most textbooks hardly mention lung tumors', and the books that did, did 'seldom get into the hands of the medical public at large, so that the general practitioner is not in a position to diagnosticate a primary lung tumor as often as might be, and the belief in the extreme rarity of these cases is still maintained'.[80] To add to these difficulties, Adler deplored that 'even the diagnoses made on the autopsy table are not always reliable'.[81] However, to Adler there seemed hardly any room for doubt 'that the increase in the percentage of lung tumors is to be attributed mainly to the increased attention paid to these types of tumor and the greater care and more extensive microscopic investigation with which autopsies are carried out at present'.[82] The observed increase in lung cancer, therefore, to him was not real, at least as far as could be concluded based on the often crude, unreliable and incomplete data that was available.

I have talked much about diagnosis in this chapter, and almost as much about post-mortem examinations. To some degree this reflects the fact that even if there had been ways of diagnosing lung cancer consistently and reliably, there was little doctors could have done to intervene. Reports on lung cancer in the nineteenth century dealt almost exclusively with patients who were killed by their disease. Unstopped by surgical or medical interventions, their cancers were left to grow until they took over, it appears, one side of the chest cavity more or less completely. Treatment was exclusively palliative and lung cancer was viewed as a disease that inevitably led to a patient's death. But in Laennec's days the same applied to tuberculosis. There were few serious illnesses that were not recalcitrant. Recalcitrance was the norm.

Even if the diagnosis was grave and not much could be done, Adler argued that a correct diagnosis was important, and that it should be made as early as possible. He was sure that a cure would come from surgery sooner or later, and that the future lay in collaboration between physicians and surgeons:

> It cannot be a matter of indifference to the unfortunate sufferer whether his case be diagnosticated as tuberculosis or as tumor. If

tuberculosis, he will be sent from one climate and one sanitarium to another, he and his family possibly deluded with false hopes, until finally secondary symptoms have cleared up the case and death has brought relief. ... At all events, so much is certain, that if the diagnosis of lung tumors is to be developed so as to render it more precise, and if any reasonable attempt is to be made to convert the present desperate prognosis into one less hopeless, this great result can only be achieved if the internist shall work hand in hand and shoulder to shoulder with the surgeon. The internist must be able to furnish as early and accurate a diagnosis as possible, so that the surgeon under favorable conditions may develop his technique as early as possible.[83]

There was not much that could be done to prolong the lives of lung cancer patients. However, this was not so different for other cancers, even those that, unlike lung cancer, were considered as surgical. Barron Lerner in his book on the history of breast cancer, for example, cites a surgeon who in 1924 confessed to a colleague that all 88 breast cancer patients on whom he had operated died subsequently from the disease.[84] Lung cancer was not a surgical disease at this stage. Patients were almost exclusively treated by internists. This was changing, however, with new developments in surgery: the emergence of radical surgical interventions inside the thorax, first to treat injuries and later also for tuberculosis, lung cancer and other diseases. This is what I will turn to in the next chapter.

3
Lungs in the Operating Theatre, circa 1900 to 1950

If there was any hope for a cure for lung cancer, Isaac Adler wrote in 1912, he expected it to come from the surgeons.[1] Lung cancer in the nineteenth century, as I have argued, had been predominantly a matter for physicians. When Adler's book was published, opening a patient's chest was still considered highly risky, if not impossible by most surgeons. This was to change over the following three decades as the human thorax became accessible to surgical intervention, thanks mostly to innovations in surgical technique and anaesthetic techno-logy. Radical surgery was becoming the treatment of choice for breast cancer, turning this disease and its treatment into a model for other cancers.[2] Surgery was associated with science and progress, and perhaps this was true especially for thoracic surgery, with its close links to tuberculosis and pulmonary physiology.[3]

Over the course of the twentieth century, the histories of lung cancer, thoracic surgery and anaesthesia, and of specialist chest services established for the treatment of tuberculous patients became closely intertwined. Until the 1950s, most chest operations were not for the treatment of lung cancer but of pulmonary tuberculosis, bronchiectasis or empyema.[4] All three diseases became manageable by other means (mostly antibiotics) after the Second World War, just when chest surgery was turning into routine. This development, as I will show in this chapter, was helped by a number of technical innovations in the late nineteenth and early twentieth century, in surgery and especially anaesthesia. Chris Lawrence has argued that technologies did not determine the course of the history of surgery.[5] Rather, the develop-ments which new inventions in anaesthesia, antisepsis and asepsis are often said to have triggered, were well under way when these technolo-

34

gies became available. Thoracic surgery was no exception: in the chapter on pulmonary resection in his *History of Thoracic Surgery*, for example, the surgeon Richard Meade lists a small number of individual cases in which operations on patients' lungs were performed before 1900.[6] Surgeons experimented with chest operations on animals since the 1880s. However, thanks to new techniques and technologies, especially during and after the First World War and in the interwar period, as I will show, pulmonary surgery became safer and more widely available. Furthermore, thanks to surgical interventions and new diagnostic techniques, such as X-ray, bronchoscopy and biopsies, larger numbers of chest patients were now diagnosed with specific lung diseases rather than assumed to suffer from consumption.

Historians have characterized surgery in the late nineteenth and early twentieth centuries as heroic, progressive, and masculinist.[7] Thoracic surgeons were no exception; they saw themselves as heroically conquering anatomical and physiological frontiers.[8] British chest surgeons were progressive-minded – not only technically but also organizationally: they sang the praises of specialization, teamwork and regionalization; and such principles were institutionalized in the 1940s with the establishment of specialist Chest Units in the Emergency Medical Service (EMS) and later in the National Health Service (NHS).[9] This chapter will take us through to the 1950s, when new modes of working pioneered in a few centres in the interwar period were implemented throughout the country. Rather than just a handful of heroic surgeons performing experimental chest operations, as was the case in the early years of the twentieth century, in 1952, according to figures compiled by the Society of Thoracic Surgeons, there were 68 consultants and 34½ senior registrar posts in thoracic surgery in the United Kingdom.[10] This was still not a vast number, but there was a good coverage through specialist chest centres. The role of surgeons changed during the first half of the twentieth century. Where before the First World War they mainly executed physicians' orders, by the 1950s chest surgeons were controlling patient pathways.[11] I will trace how radical surgical operations such as lobectomies (the removal of the affected lobe or lobes of a lung) or pneumonectomies (the resection of a whole lung) turned from pioneering operations used predominantly to treat bronchiectasis, into standard treatment methods for lung cancer, then still considered rare. I will concentrate on the history of thoracic surgery in Britain and complement the story with references to developments in the United States and on the European continent.

Technique, technology and surgical specialization

While the Brompton Hospital for Consumptives and Diseases of the Chest employed consultant surgeons from its inception in 1842, their work initially consisted in general surgical support rather than chest surgery. Towards the end of the nineteenth century, however, a new set of specialist surgical interventions came into use for the treatment of pulmonary tuberculosis. 'Collapse therapy' gave surgery an increasingly central role in the treatment of this and other diseases of the chest.[12] The surgical therapy of tuberculosis ranged from the Artificial Pneumothorax (A.P.) to thoracoplasty, the attempt to relax the affected lung by removing one or several ribs and thereby reducing the size of the thoracic cavity.[13] These operations were progressive, interventionist, and based on ideas of scientific surgery. The intention of both operations was to put the diseased lung 'at rest'. The spread of disease was to be prevented by restricting the lung's movement. In the case of A.P., the older of the two procedures, this was done by way of introducing air or another gas into the chest, causing the lung to collapse. The method was first applied by an Italian, Carlo Forlanini in 1882, publicized in the US and Germany just before the turn of the century and introduced to Britain around 1910. It was considered so simple a procedure that it could be performed by physicians as well as surgeons. The Brompton Hospital's Consulting Physician from 1905 to 1934, Sir James Kingston Fowler, described A.P. as 'the only advance in the treatment of pulmonary tuberculosis since the introduction of sanatorium treatment'.[14] The establishment of sanatoria is also important in this context, as they provided the institutional context for much thoracic surgery. What makes the routine use of such procedures in the treatment of tuberculosis relevant to a book on the history of lung cancer is that, along with the treatment of war casualties, they paved the way for other surgical interventions inside the chest cavity.

While the new techniques were introduced earlier on the European continent, surgery was still marginal at British chest hospitals at the eve of the First World War. At the Brompton Hospital, during the whole year of 1914, only 61 surgical operations were carried out, and these were classed as general surgery rather than specialist chest operations and could have been performed in any general hospital.[15] While operations on the abdomen had become possible thanks to anaesthesia and the application of the principles of antisepsis and asepsis, most general surgeons – and there were few specialists – still viewed opening the chest cavity as a taboo. Some, however, like Hugh Morriston Davies at the London Chest Hospital (about whom we hear more later), increasingly considered operations on the lung a possibility. The main problem that surgeons operating inside the chest were facing was the so-called 'pneumothorax problem'.[16] If the thorax was opened, the lung on the affected side was at danger of collapsing. If this occurred – a so-called open pneumothorax – and the patient attempted to breathe, the (used) air, rather than being exhaled and exchanged for fresh air, was transferred back and forth between the collapsed and the healthy lung. It is important to note that surgeons had to rely on patients' own, spontaneous breathing during operations until the late 1930s, when methods of manually assisted respiration were promoted by some anaesthetists and the first commercially developed mechanical respirators became available. A patient affected by an open pneumothorax ran out of oxygen fairly quickly, leaving a surgeon operating on the lung very little time to finish his work.

An early technical solution to the pneumothorax problem was proposed by the German surgeon Ferdinand Sauerbruch. The negative pressure chamber (Figure 3.1), presented in 1904, came to constitute something like an origin myth of modern thoracic surgery. Sauerbruch's idea was to enclose both the chest of the patient and the

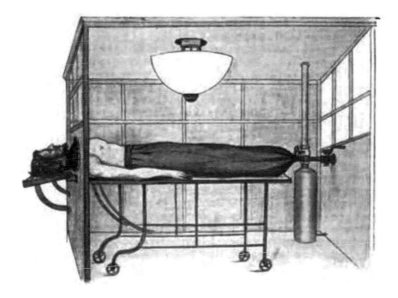

Figure 3.1 Sauerbruch's Negative Pressure Operating Chamber.[17]

surgeon in an airtight chamber, where pressure was kept lower than on the outside. The head of the patient was outside the chamber, with the effect that the higher air pressure in the lungs kept them from collapsing when the chest was opened. The patient, meanwhile, continued to breathe spontaneously. Sauerbruch developed this device with his senior at Breslau, Johannes von Mikulicz-Radecki.[18] However impressive, the device was cumbersome and expensive and only a few were built. Other surgeons developed different solutions. Morriston Davies, for example, designed a machine in 1911 that worked by applying increased air pressure directly to a patient's lung through a tube inserted into the trachea (Figure 3.2). This was not a respirator, however, and the patient still had to breathe spontaneously.[19] The machine was more mobile and less cumbersome for surgeons, and similar positive pressure devices, designed and further developed by

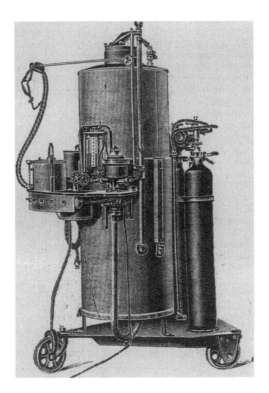

Figure 3.2 Morriston Davies's Positive Pressure Anaesthetic Machine.[20] Reproduced with permission.

others, found wider application in chest surgery after the First World War. Morriston Davies used his machine when in 1912 he performed what is viewed by many as the first successful dissection lobectomy (the removal of one or several lobes of a lung), apparently using a technique almost identical to procedures used later in the century to operate on cancerous lungs.[21] The lobectomy technique which Morriston Davies used was not to be applied again until the late 1920s, mostly because chest operations were usually not undertaken for cancer but for chronic, infectious conditions such as bronchiectasis, which led to parts of the inflamed lung tissue attaching itself to surrounding tissues, making this type of operation too time-consuming to manage patients with the devices and gases that anaesthetists then had at their disposal. Speed was still essential for intrathoracic procedures. Morriston Davies' patient, incidentally, died eight days after the operation.

The 1912 lobectomy was one of 11 'firsts' that Morriston Davies was credited with between 1911 and 1914.[22] Originally set to become a neurosurgeon, Morriston Davies had attended a conference in Germany in 1910, where Sauerbruch and others presented papers on their groundbreaking chest operations using negative pressure chambers. Impressed by what he saw, Morriston Davies decided to turn to chest rather than neurosurgery. Like his friend, the neurosurgeon Wilfred Trotter and many US pioneers of thoracic surgery, and unlike many other contemporary British surgeons, he remained interested in academic research. Back from the conference, Morriston Davies had an X-ray apparatus installed in his research rooms and experimented with using the new equipment for the detection of chest diseases. During this period he also designed his pressurized breathing apparatus and other devices. In fact, much of the article in which he reported the 1912 lobectomy (as one of several cases, treated for different chest conditions), was concerned with the use of X-rays in the diagnosis and of various new devices in the treatment of chest disease. Chest physicians were critical of Morriston Davies's experimental approach to treating chest disease.[23] Nevertheless, in 1914 Morriston Davies was promoted to full surgeon and appointed to the staff of the London Chest Hospital.

An accident in 1916, which changed Morriston Davies's career path and nearly cost him his life, demonstrates that surgery before the introduction of antibiotics carried significant risks not only for patients but also for surgeons.[24] During an operation Morriston Davies injured his right thumb and developed septicaemia which spread to his arm. He survived, the arm did not have to be amputated, but the surgeon lost the full use of his right hand. It was not clear if he was ever going to be able to operate again, and he had to reconsider his career plans. In 1918 he accepted an offer to run a private, 30-bed sanatorium, Llanbedr Hall in North Wales. With the help of special instruments which he designed himself, he even learned to operate with his left hand.[25] He built an operating theatre at Llanbedr Hall and convinced many of his patients of the advantages of surgical treatment. In 1920 he became Consulting Thoracic Surgeon to two sanatoria run by the King Edward VII Welsh National Association, at Llangwyfan and Talgarth, and in 1925 he also accepted an appointment at the Cheshire Joint Sanatorium. From 1932, in addition to his other appointments, he also co-ordinated the development of surgery at sanatoria in East Lancashire. Morriston Davies managed to combine these jobs by driving long distances in his car.[26]

Apparatuses such as those developed by the surgeons Sauerbruch in Germany, Morriston Davies in Britain or Samuel Meltzer in the United States, did much to make chest surgery more practicable. As was the case for other fields of surgery, thoracic surgeons made their names as much with technical innovations as with publications. But devices and techniques developed by anaesthetists were just as important. Ivan Magill, for example, consultant anaesthetist to the Brompton Hospital from 1921 to 1950, worked closely with the hospital's surgeons during this phase of rapid expansion. The increasingly important role played by Magill or fellow anaesthetist Michael Nosworthy at St Thomas's Hospital, illustrates the shift of focus away from the lone, heroic surgeon to the surgical team. Magill developed new methods of intubation and artificial ventilation that were to become essential to modern chest surgery, as they allowed the selective ventilation of one lung, while the other was operated on. Magill's apparatuses were small, portable, and relatively cheap (unlike Sauerbruch's negative pressure chamber), making them affordable also to smaller, provincial hospitals.[27]

Devices and gases: The role of anaesthesia

Looking at Ivan Magill's career in anaesthesia, an area of medicine that became increasingly more differentiated during this time of specialization in surgery, allows us to illustrate the importance of developments in this field. Magill's work in anaesthesia began after the First World War when, according to his own admission 'more by chance than by choice', he was posted as anaesthetist to the Queen's Hospital, Sidcup.[28] The patients treated at Sidcup were war casualties with wounds to face and jaws. When these patients underwent surgery, applying anaesthesia by conventional means, as Magill remembers, led to a 'constant struggle' to keep the patient's airways free without obstructing the surgeon or encroaching on the aseptic field.[29] Together with his colleague Stanley Rowbotham, Magill started to insufflate patients by way of an elastic rubber catheter that they inserted into a patient's trachea, through which air was driven by a motor pump. Later they also introduced a second tube for expiration, which prevented surgeons from being exposed to ether laden expirations, sometimes accompanied by a spray of blood. The tubes were first introduced through the mouth and subsequently through the nose, which kept them more safely outside the aseptic field. The 'catheters' were not produced for this purpose but purchased from a shopkeeper who dealt in rubber tubing of all sizes. Magill selected pieces that had a natural curve because they were stored in coils. Endotracheal intubation (inserting the tube deep into the trachea) proved useful during chest operations such as lobectomies or pneumonectomies, either to allow the application of a mix of air with an anaesthetic agent such as chloroform, nitrous oxide or by the 1930s cyclopropane (a gas that could be administered in a mixture with a higher proportion of oxygen, making it less likely that the patient suffered oxygen shortages), or just to keep the airways free of secretions by using the tube as a suction device. Sometimes local or spinal anaesthesia was applied, and depending on the case and the preferences of anaesthetist or surgeon, the tube was inserted all the way down through the trachea into either the affected or the unaffected bronchus, and fixed there with an inflatable cuff.[30] While there was still some controversy in the 1930s as to whether intubation was really necessary in chest operations, by the 1940s anaesthetists increasingly agreed that one-lung anaesthesia by way of inserting the tube into one of the bronchi was ideal for pneumonectomies.[31] During the Second World War, the new technologies became routine, and they were firmly established as part of the toolkit of chest surgery teams by the end of the war.

Opening the thorax

The new devices and approaches in anaesthesia that made routine chest surgery feasible were only widely introduced after the First World War, but it was the war that confronted many surgeons with chest injuries of a new kind and made it almost unavoidable that some gained experience with chest operations.[32] Before 1914, the chest wounds that military surgeons encountered were generally caused either by a knife or by low-velocity bullets. Surgeons were ill prepared for the much more extensive wounds caused by the use of high-explosive artillery and high-velocity rifle bullets on a large scale. Furthermore, during the great influenza epidemic of 1918, operations for empyema (not a direct symptom of the flu but a consequence of frequently occurring secondary streptococcus infections) were often interventions of last resort when treating flu victims, both soldiers and civilians. Their chests filled with fluid, making breathing more and more difficult. Opening their chests surgically to drain away this pus was often their last hope. What made things more complicated was that the fluid accumulating in the pleural space of these influenza patients was less viscous and sticky than in other forms of empyema, making the collapse of the lung more likely when the chest was opened. The treatment of flu victims under these difficult circumstances exposed many surgeons to emergency chest operations for the first time. One of the American pioneers of lung cancer surgery, Evarts Graham, became a chest specialist partly as a consequence of his involvement with the Empyema Commission, which was established by the US Surgeon General in 1918 to investigate the best way of treating this condition and thus address one of the most feared and deadly effects of the flu epidemic.[33]

The opening of the thorax with its vital organs was (and probably still is) an awe inducing operation, daring and with great risks attached. The graphic, almost visceral description of an attempted lobectomy for bronchiectasis by the American surgeon Samuel Robinson included in his presidential address to the American Association for Thoracic Surgery in Washington in 1922, may convey some of the tension, the surge of adrenaline associated with operations of this kind, where a surgeon often worked on the contents of the chest without seeing what he was doing, trying to find the right place to separate the diseased parts of the affected lung – filled with pus – from those that were healthy; under his hands a patient whose condition was becoming rapidly more distressing. Robinson tells, in his own words, the 'story of a more or less typical operation for bronchiectasis to emphasize the difficulties which are facing us in lung surgery'.[34] The confidence that many surgeons had acquired when dealing with war injuries of the chest, he argued, was treacherous when it came to operations of this kind:

> The enthusiast returns from the war who has often dragged a lung
> lobe into a spread thoracic wound, opened it, scraped it, washed it,

yea, even removed it – and concluded, therefrom, that intrathoracic surgery is freed of its supposed dangers, and that the possibilities therein are comparable to those in the abdomen. Let him attempt the same performance in his hospital in a case of lower lobe bronchiectasis. Then he will learn what real thoracic pathology means. The patient is placed on the operating table. The posture is uncomfortable. There may be cyanosis. It induces coughing. The anesthetist is greeted by an evacuation of a large amount of pungent, purulent sputum, incident to the posture on the table. The whole bronchial tree may be filled with this material as the anesthetist begins. ... As the secretions well toward the trachea, the cyanosis increases. The lower lobe obstinately resists being delivered; the pleural adhesions are strong and widespread; the attachments to the diaphragm are ropelike and tenacious. Finger dissection is inadequate. Work with the knife and scissors is blind. Cleavages are sought in vain. ... Meanwhile, the patient's condition may become distressing and perhaps alarming. If open pneumothorax is adding insult to injury, the lung cannot be used to plug the thoracic gap, because the lobe is not deliverable.[35]

The use of one of the new positive or negative pressure machines designed to overcome the pneumothorax problem sometimes led to new, unexpected difficulties, confronting the operating surgeon with the decision if he could continue with the operation or needed to abandon it to rescue his patient:

There may be cyanosis, even with the head outside a negative pressure cabinet. And then the difficulties multiply. The complete liberating [i.e. removal of the affected tissue – CT] at one sitting may have to be abandoned. There is bleeding and infectious leakage from the lung, and bleeding from the diaphragm. Tight closure of the chest without drainage seems inadvisable under such conditions, and yet necessary to avoid the ills of postoperative pneumothorax. Suddenly, it is obviously time to return the patient to his bed. Not much has been accomplished.[36]

How did surgeons justify exposing their patients to such operations, which obviously came with serious dangers attached? By pointing out that the disease was much worse than the cure: bronchiectasis patients usually died young, after a long period of suffering. As is the case with lung cancer, many deaths from this condition were probably never attributed to bronchiectasis, so it is difficult to estimate how common it was. An often reprinted quote by an American pioneer of thoracic surgery, the professor at Cornell University, Howard Lilienthal, illustrates the distressing nature of the condition as it presented itself to surgeons:

> Occasionally an individual coughs his way through life – never a long one – and manages to exist as a semi-invalid, with copious, foul expectoration which no medicine can control, being a handicap difficult to bear. Patients have even threatened suicide if refused the chance for cure by operation, though they knew that the danger was great.[37]

Lilienthal performed his first lobectomy to treat bronchiectasis in 1914, and 31 by 1922, some of them in two or several stages rather than as a one-stage operation such as that performed by Morriston Davies on a lung cancer patient two years earlier.

Lilienthal recommended an X-ray examination to clarify extent and location of the disease. Bronchoscopy, the examination of the inside of the bronchus through an inserted, rigid metal tube that contained a light source and a set of lenses and prisms, was another diagnostic technique that was usually employed in these cases.[38] This was not always necessary, however, Lilienthal argued, as the chest was going to be opened anyway, and opened widely, which would reveal conditions that could not have been predicted by any other means. Usually Lilienthal made a long incision on the patient's back, along the seventh rib. He then removed small pieces of one or several ribs and opened the thorax with an instrument called a rib-spreader. With the thorax opened, he examined the lung to get an idea of the extent to which it was affected and how much adhesion there was between the lung and surrounding tissues. He tied silk thread around affected parts of the lung, which he subsequently cut away. Ideally the whole procedure was completed in one stage, but in some cases it took two operations. Finally he placed iodoformized gauze on the stump which would later be removed through an opening in the rib cage, and introduced a drainage tube lower down in the chest. Some patients lost considerable volumes of blood and suffered severe shocks, which were treated with blood transfusions. Others died on the operating table or shortly after surgery. Operative mortality was high: by 1922, six of the 14 patients where Lilienthal removed one lobe of a lung, did not survive the intervention (this corresponds to a mortality of 43 percent). Of the ten patients where he removed more than one lobe, seven died (70 percent mortality).[39] Lilienthal emphasized that he always fully explained the operative risks to patients and their relatives before accepting cases for operation: 'They should request me to operate; I do not try to persuade them'.[40]

Lung cancer under the knife

While lung cancer was not considered common, some lung cancer patients did find themselves on the operating tables of thoracic surgeons, besides those suffering from tuberculosis or bronchiectasis. In 1912 an unnamed man, aged 44 was referred to Hugh Morriston Davies at University College Hospital London because he was suffering from what had been diagnosed as a persistent case of bronchitis and emphysema, possibly suitable for a new type of surgical operation. His bronchitis had been worse over the winter and he had also felt pain in the right side of his chest for the previous four months. As part of the routine investigation the patient underwent a radiological examination, which revealed a shadow on the lung, indicative of a tumour. Over the next three weeks the patient was repeatedly examined by several experts, but no physical signs other than those of bronchitis and emphysema could be discovered (this illustrates how difficult it was to diagnose a tumour based alone on clinical signs). A week after the X-ray examination the patient started to bring up sputum that looked like prune juice and contained a type of cells that Adler had described in his recently published book as indicative of lung cancer. Morriston Davies was keen to operate (not least, probably, to test his X-ray diagnosis), but the patient refused the operation for another two months. Finally he consented. He was anaesthetized with ether, a tube was inserted into his trachea, and he was connected to Morriston Davies's positive pressure apparatus to prevent pneumothorax. The surgeon opened his chest with a long cut along the sixth rib. He examined the lung, presumably relying on touch, and found that the tumour was confined to one lobe, with some attachment to the chest wall in one place. He tied a silk thread around the tissue connecting the affected lobe to the rest of the lung and removed it, along with the affected portion of the pleura. Then he stitched over the open end of the bronchus, covered it with adjacent lung tissue, and closed the chest again. We have no way of knowing how exactly the patient felt when he awoke; all Morriston Davies tells us is that his condition was 'quite good for the first six days; however, he then developed an empyema, and died on the eighth day'.[41]

Would it have been better for the patient to stick with his original decision and refuse the operation? One wonders. He was certainly going to die, but he would have gained time. To Morriston Davies, however, the operation was a success, as the autopsy showed that the bronchial stump had been healing well. The authors of insider histories of thoracic surgery agree with him, as the techniques he used were much more like more recent, 'modern' approaches to lung resection than, for example, the rather crude methods applied in the 1920s by Robinson or Lilienthal.[42] They hail him also for being ahead of his time with his attitude towards the treatment of lung cancer: if in doubt, operate. 'Cancer of the lung', Morriston Davies wrote in his article,

> is in some of its varieties, and in its earlier stages, now accessible to surgical intervention, and complete removal; but until this fact is more fully recognized, and all pulmonary cases are subjected to routine radiological examination, the growths will not be recognized until they have extended beyond the possibility of all treatment. In all doubtful cases, at least an exploratory thoracotomy should be undertaken.[43]

Outside specialist centres, lung resections for new growths, benign or malignant, were rarely undertaken in Britain until well into the 1930s. This was not only due to operative technique or problems with anaesthesia, but also diagnosis and referral patterns. The Newcastle surgeon George Mason observed in 1936 that, while resecting a lung affected by a tumour was a good idea, in practice growths had usually 'progressed beyond the bounds of surgical enterprise when first seen by the thoracic surgeon'.[44] General practitioners tended not to consider lung cancer as a diagnosis in patients that came to them with non-specific symptoms of respiratory illness – which they treated essentially like consumption. Why would they do otherwise? When they trained they learned that lung cancer was a very rare disease. And only exploratory surgery could deliver a reliable diagnosis. In any case, lung cancer was not considered treatable.[45] Indeed, most of the small number of patients who did undergo operations for lung cancer died either during surgery or shortly afterwards. The first fully successful operation for lung cancer is credited to the American surgeon Evarts Graham, described by Meade in his *History of Thoracic Surgery* as the 'most dramatic contribution to pulmonary resection' and ultimately turning Graham into a surgical legend.[46]

Evarts Graham, the surgeon credited with first performing, in 1933, the type of operation that millions of lung cancer sufferers would undergo from the 1940s onwards, used a technique that Hurt in his *History of Cardiothoracic Surgery* characterizes as 'remarkably crude'.[47] Graham performed two-stage lung resections, first opening the chest and packing it with iodoform gauze, and several days or weeks later opening it for a second time, to sink a large, red-hot soldering iron into the opened thoracic cavity of his patients, burning away the affected sections of the lung. Graham used this method, which resulted in considerable smoke and stench in the operating theatre, in order to address one of the main dangers to a patient's life following a lung resection: before antibiotics were available, in many cases the gaping wounds became infected. Even if the patients survived the days after the operation, the thoracic stumps often did not heal, forming a connection between the outside and the pleural space (known as a bronchial fistula). The risk was especially high when the lungs were filled with infectious pus, one of the defining aspects of bronchiectasis. Burning the tissue away, Graham hoped, would help reduce the risk of infection. Indeed, survival rates improved.[48] By 1930, in fact, the operative mortality for these operations went down to 11 percent.

A step more radical than the lobectomy was the pneumonectomy, the complete removal of a lung, first reported in the early 1930s. Like lobectomies, these were crude, heroic operations. Take for example the first pneumonectomy in Britain, performed in 1934 by the Newcastle surgeon George Mason, assisted by Laurence O'Shaughnessy and Andrew Logan.[49] The patient, as in many of these cases, was young, a 15-year old boy with bronchiectasis. Logan remembered in 1986 that during the operation the boy occasionally made convulsive leaps because he was incompletely anaesthetized. The surgeons opened the chest with a long incision, almost half way around the body from spine to sternum. They 'mobilized' the lung with some force as it was attached to the chest wall in many places due to the infection, then tied a rubber catheter around the stem of the bronchus, surrounded the lung with gauze, and closed the chest wall again. On the following day Mason left for a ski holiday in Switzerland. The patient became increasingly ill, but the amount of sputum he coughed up decreased. Ten days after the original operation Logan re-opened the boy's chest and removed the gauze packs, along with a large quantity of pus. The lung was obviously necrotic (which was expected and the desired outcome) and Logan cut it off, leaving the bronchus open. He left the chest wide open for another three months, during which the large, empty, pleural space was daily 'mopped' clear of pus. The space became gradually smaller, until it finally disappeared and the fistula from the bronchus connected directly to the surface of the chest. After five months, when the fistula had still not closed, the surgeons removed some ribs to reduce the residual space. The boy finally recovered.[50]

A pneumonectomy for lung cancer was the operation that secured Evarts Graham entry into the surgical hall of fame. On 27 February 1933, Dr James Gilmore, a 49-year old obstetrician from Pittsburgh was admitted to Graham's chest clinic at Barnes Hospital, accompanied by his referring physician. Gilmore had been diagnosed with pneumonia in 1929, from which he recovered. In July 1932 he suffered general discomfort, chills and fever. Blood tests revealed an elevated white blood cell count. A chest X-ray showed a fan-shaped shadow in the upper lobe of his left lung. The condition regressed but recurred. An attempted aspiration of what was then suspected to be a lung abscess, went wrong and led to a pneumothorax that persisted until Gilmore's referral to Graham's clinic. A chest radiograph taken at Barnes Hospital revealed that the affected lobe had collapsed, a bronchogram showed that there was an obstruction, but a bronchoscopy and biopsy revealed no clear findings. Two further visits to the chest clinic followed in March, including another bronchoscopy and removal of a tissue sample for a biopsy. The surgeons feared the sample might be insufficient for diagnosis, but the pathologist for the Ear, Nose and Throat Service reported a squamous cell carcinoma. A right upper lobectomy was scheduled for 5 April. Gilmore returned to Pittsburgh and went to his dentist to have some fillings replaced (signalling hope and confidence) and purchase a cemetery plot (signalling perhaps realism).[51]

Gilmore returned to Barnes Hospital on 4 April and arrived in the operating theatre shortly after nine the following morning. He was anaesthetized with nitrous oxide and oxygen, and a Magill tube was inserted into his trachea. Graham and his assistant removed two ribs and opened the chest. When examining the lung with his hands Graham found a mass that involved the main stem of the bronchus and the whole of the upper lobe. Gilmore was accompanied by his brother-in-law, a physician and another physician friend. Graham told them that he considered a lobectomy useless and advised to remove the whole lung. Alas he had only performed this operation, a one-stage pneumonectomy, in animal experiments. After some discussion, he decided to go ahead. The surgeons placed a rubber catheter around the stem of the lung to constrict the blood flow, added two metal clamps, cut between them, removed the lung with one clamp, placed three sutures around the stump, and removed the other clamp. Combining chest surgery with radiotherapy, they cauterized the stump with heat and silver nitrate and implanted radon seeds to irradiate tumour cells that may have been left in the chest cavity. Then they removed seven more ribs to reduce the size of the empty chest cavity and allow the chest wall to collapse onto the stump.[52]

Gilmore's operation was an exceptional success, as Graham realized pretty quickly. Not only did his patient survive the hours and days after the operation, he made a full recovery and travelled to conferences with Graham for show and tell sessions. Gilmore ultimately even survived his surgeon: Graham, a habitual cigarette smoker, died himself from lung cancer in 1957. Gilmore, incidentally, also continued to smoke. But the success was exceptional in different ways, too. The operation on Gilmore was followed by a series of failures. Persistent rumours circulated in St Louis that the next 16 operations all ended in fatalities, and Graham became known as the 'Butcher of Barnes Hospital'.[53] C. Barber Mueller in his biography of Graham suggests that these deaths included bronchiectasis patients and that the mortality figure for the (still rare) lung cancer operations was lower. Still, the operation was far from safe.

Others made lung resection safer and so enabled Graham to become a legend within another decade or two. William Rienhoff was professor of surgery at Johns Hopkins Hospital, Baltimore. He developed what would become the established technique of dissection pneumonectomy by applying what Raymond Hurt in his *History of Cardiothoracic Surgery* calls 'general surgical principles' to the procedure.[54] His approach was minimalist; he advised against removing ribs or cauterization, stressing that the most important factor was to keep blood circulation to the bronchial stump intact to ensure good healing. He also operated remarkably fast. The other important standardizer of pneumonectomy was Clarence Crafoord, a Swedish surgeon. He published a highly influential monograph in 1938, *On the Technique of Pneumonectomy in Man*.[55] Crafoord emphasized the importance of good anaesthesia, stating clearly that he felt positive pressure anaesthesia was far superior to Sauerbruch's negative pressure approach.

Surgery at a British chest hospital

However heroic and dangerous these semi-experimental procedures may have been, the Brompton Hospital in London saw a marked increase in the number of chest operations (including a growing number of lobectomies) after the end of the First World War and following the appointment of James Ernest Helme Roberts (commonly referred to as J. E. H. Roberts) in 1919 and Arthur Tudor Edwards in 1922.[56] Both became well-known thoracic surgeons, and Tudor Edwards turned into a figure head of the new discipline beyond the borders of the United Kingdom. He was known as 'an operator of supreme skill and beautiful technique' and is credited with the first successful one-stage dissection lobectomy in 1928.[57] Priority claims like this one are popular in the history of surgery, and sometimes there are several firsts for a particular procedure. Did not Morriston Davies perform the first successful operation of this type? Morriston Davies's patient in 1912, however, died after eight days, and his was a lung cancer patient – then still considered a very rare disease – rather than a bronchiectasis sufferer. The case was buried among other cases in a long article reporting the use of various new techniques in the diagnosis and treatment of chest disease, and it is likely that it was overlooked. Tudor Edwards was also the first in Britain to perform a successful one-stage pneumonectomy (the removal of a whole lung in one operation) in 1935, followed two days later by his colleague and rival Roberts. Due to innovations in anaesthesia, as discussed above, lobectomies and pneumonectomies became routine after the Second World War era; and these heroic operations came to define the specialty.[58] According to Tudor Edwards's obituary: 'His reputation was established through his pioneer work in developing techniques which helped to advance thoracic surgery from the occasional reluctant, and always precarious, intervention, to the status of an acknowledged specialty ranking with abdominal and other accepted branches of surgery.'[59]

The Brompton Hospital dealt predominantly with suspected cases of tuberculosis until after the Second World War, but employing and developing the diagnostic techniques that were later also used routinely in the detection of lung cancer cases. The hospital was changing from an institution dedicated to care for incurables to a site of technological innovation adopting a more interventionist ethos and treating more patients (not unlike cancer hospitals in the same period). In 1900 a Radiological Department was set up, where 40 diagnostic examinations were undertaken in the first year. By 1907 the number reached 168, of which 45 were cases of tuberculosis. Only in 1919 the department was open daily. The number of examinations per year by then was 1,191. By 1932 the Brompton Hospital undertook 10,000 X-ray examinations, by 1938 20,000 and by 1948 35,000. The number of bronchoscopies, an important diagnostic procedure usually performed by surgeons also increased, albeit more slowly. Only one bronchoscopy was performed at the hospital between 1908 and 1925. Over the next four years, there were 29, and in 1930, 28 were performed. By the early 1950s, the number went up to 800 per year.[60]

A pathology department was established in 1906. Before this date, as was generally the case in Britain in the nineteenth century, the physicians were in charge of pathology.[61] The pathological work then comprised chiefly autopsies. Also in 1906 a Clinical Laboratory was set up, which among other tasks was in charge of the routine examination of sputum.[62] The pathologist would also examine the increasing numbers of biopsy samples taken from lung cancer patients during operations. The figures point to a rapid expansion in the number of examinations and diagnostic procedures. But how were patients at the Brompton Hospital and other chest hospitals treated?

The imminent rise of thoracic surgery was not obvious in the early 1920s. Maurice Davidson and Frederick George Rouvray in their history of the Brompton Hospital characterized the work of Tudor Edwards and Roberts in the interwar period, as 'an uphill task'.[63] There were no separate surgical beds available for them at the hospital; their surgical technique was new and not standardized, and they had to work out much of it by themselves, to a large extent by trial and error. We can only speculate what this meant for patients. Roberts and Tudor Edwards found themselves confronted, according to Davidson and Rouvray, by an 'atmosphere of doubt and apprehension'.[64] Considering the risks associated with some of the operations they were carrying out, this was perhaps not surprising.

The numbers of surgical operations performed at the hospital in the 1920s were initially roughly the same as before the war. But more operations than before the war were specialist chest operations rather than general surgery: thoracoplasties, explorations of the chest or drainage operations.[65] In 1921, out of a total of 66 operations, the majority involved the chest.[66] In 1922 Roberts and Tudor Edwards operated in 78 cases, and again, most of the operations were thoracic. 'There has been a decided increase in the surgical work of the hospital', the surgeons commented in the medical report: 'It will be remarked that there is an increasing tendency to treatment by surgical measures.'[67] The number of operations performed per year continued to increase. In 1926 the surgeons acquired their own wards and a new operating theatre was built. With the new wards and theatre, the number of operations increased further. In 1931, Roberts and Tudor Edwards operated on 603 cases, almost ten times as often as a decade earlier. In 1933, 1,122 operations were performed. But not only did the absolute numbers of operations increase: figures which Russell Brock compiled in 1964 (a member of the next generation of thoracic surgeons and a postwar successor to Tudor Edwards and Roberts), based on the operating books of the Brompton, show that around 1920 less than 50 percent of the operations were thoracic. By 1928 chest surgery accounted for over 70 percent of all operations, and by the mid 1930s, more than 95 percent of operations were thoracic.[68] The total numbers and the ratio of normal to chest surgery were to remain roughly the same until the 1960s. In 1963, out of a total of 1,289 operations, 1,268 were thoracic.[69]

In an article on the history of his discipline, the thoracic surgeon Roger Abbey Smith identified relative ignorance about the anatomy and physiology of the lung, inadequate anaesthesia, uncontrollable sepsis, and fear of the open pneumothorax as the main problems facing chest surgeons in the 1920s.[70] In his opinion the fear of the open pneumothorax was greatly exaggerated. The risk of complications was real, however, and as a consequence, post-operative mortality and morbidity rates were high. The incidence of broncho-pleural fistulae after lobectomies, for example, was 30 percent in the late 1930s.[71] From the 1940s onwards antibiotics helped to reduce such risks. Outside special hospitals and chest sanatoria, there were additional difficulties that Abbey Smith blamed for making thoracic surgeons less effective, such as the absence of experimental animals that allowed surgeons to experiment with new operations, the lack of organization and the unwillingness among medical staff to develop team-oriented approaches.

Abbey Smith argues that not only did the surgeons have to be specialists to secure good results, so did nursing staff. Calls for teamwork in medicine and surgery, a twentieth century idea, became more frequent in the interwar period, as Roger Cooter has suggested, 'deployed to criticise allegedly uncoordinated, haphazard, and inefficient medical arrangements'.[72] When thoracic surgeons published on the history of their discipline, they often emphasized the importance of such arrangements (besides technical innovations). Tudor Edwards's obituary in the *Lancet* describes the pioneer thoracic surgeon as 'an outstanding organiser of team-work'.[73] The team in this case included not only staff, but also, apparently, the patient:

Surgeon, physician, radiologist, pathologist, anaesthetist, physiotherapist, nursing staff, surgical assistant – all knew what was expected of them and gave their best. The patient was also made fully aware of the nature of the operation planned, of its risks, and of any disabilities it would entail. His full cooperation was obtained in a carefully planned course of preoperative and postoperative treatment.[74]

Collaborations also extended beyond the doors of the hospital, especially where cancer was concerned. In 1944 the Brompton Hospital was formally approached by the Royal Cancer Hospital with the proposal to set up a joint clinic, employing specialists from both hospitals, to explore the possibilities of X-ray therapy in cases of lung cancer where surgery was deemed impossible or undesirable. I will discuss the work of this joint clinic in Chapter 5.

British chest surgeons getting organized

One factor changing general surgeons performing predominantly chest operations into self conscious thoracic specialists, was the emergence of societies and clubs specifically for thoracic surgeons. In Britain, the first of these associations was the Society of Thoracic Surgeons, the precursor of today's Society for Cardiothoracic Surgery in Great Britain and Ireland. The Society was launched following an initiative of the Manchester surgeon Alexander Graham Bryce in 1931.[75] The inaugural meeting took place on Friday, 5 May 1933 at the Midland Hotel in Manchester. Morriston Davies was the Society's first President; and Roberts and Tudor Edwards were Vice-Presidents.[76] The Society was small, more like a club than a large professional association (the initial plan was to only admit up to 25 members). It held annual meetings, either hosted by one if its members in Britain or going abroad (for example in Davos or at Sauerbruch's clinic in Berlin). Meetings involved presenting papers as well as, possibly more importantly, watching other surgeons performing chest operations.

The somewhat smaller British association was modelled on the American Association for Thoracic Surgery (AATS), then led by Evarts A. Graham. The AATS had been launched a few years earlier, in 1918 by Willy Meyer, president of the New York Association for Thoracic Surgery. From its inception, there were some members more interested in respiratory physiology, others more in practical surgery (and many combined these interests). Some among the British surgeons thought that the American Society valued experimental science too highly, while clinical matters did not receive enough attention.[77] The first generation of British thoracic surgeons were, in fact, more craft than science-oriented (with the possible exception of Morriston Davies); Tudor Edwards, for example, published relatively few articles. The early AATS, in contrast, was informed by continental European traditions. Its first president, Meyer's friend Samuel J. Meltzer (1851–1920) was born in Russia into a Jewish family, went to school in Königsberg (East Prussia), and trained in medicine at the University of Berlin.[78] In 1883 he emigrated to the US, to New York City, where he developed a busy medical practice, while at the same time pursuing his interests in experimental physiology. He was a founder and first president of the Society for Experimental Biology and Medicine. In 1904 he accepted the offer to spend half of his time as a physiologist at the Rockefeller Institute, and within three years gave up his practice completely. In 1909, along with his son-in-law, he developed an oxygen insufflation technique via an intratracheal tube as an alternative to Sauerbruch's negative pressure chamber. According to the author of a biographical sketch in the *Journal of Thoracic Surgery*, Meltzer 'astutely recognized the special promise of thoracic surgery if developed in company not only with chest physicians, tuberculosis specialists, and anaesthesiologists, but also with basic scientists'.[79] Evarts Graham, president of the AATS in 1928 and editor of the *Journal of Thoracic Surgery* from 1931 until his death in 1957, was similarly experimentally-minded. He valued experiments and laboratory research, but like traditional surgeons he built his reputation on heroic operations. In Graham's case, the most famous operation he is credited with is the first successful pneumonectomy for carcinoma of the lung, performed in 1933.

Utopian visions for British chest surgery services

In 1938 the Society of Thoracic Surgeons started a campaign for a regional reorganization of chest services, with centres staffed by specialists.[80] In 1941, a circular was distributed, asking members to report what facilities for chest surgery were available in their area, in all hospitals they knew about (including those established under the Emergency Medical Service). Based on the replies, a Memorandum on the Provision of a National Thoracic Surgery Service was drafted.[81] Drafts were exchanged, discussed and amended, until, finally, in 1944, the Memorandum was published. Three hundred copies were printed and distributed to members of the Society, Royal Colleges, medical schools and health officials. A slightly revised version was published in March 1948.[82]

The memorandum depicted a somewhat Utopian scheme for a regional service. The time had passed, a *BMJ* author summarizing the memorandum argued, when physicians directed a general surgeon to perform some particular operation on the chest.[83] If such surgery was to be done well and advances were to be achieved, thoracic surgeons had to be well trained, and they had to be able to rely on the cooperation of many associated workers. High degrees of surgical and anaesthetic skills were required, and proper pre- and post-operative management depended on the availability of specially trained nursing staff and physiotherapists. The thoracic surgeon had to be backed up by a 'proper organization or team'.[84] What was needed was a properly organized service with centres adequately staffed and equipped.

The authors of the memorandum dedicated some space to the issue of teaching thoracic surgery. All surgeons should receive some teaching in the subject, enough to familiarize them with diagnostic methods and the type of surgical help that could be offered to chest patients. For this purpose, undergraduate teaching hospitals needed thoracic surgery departments with some 25 beds and an outpatient clinic. Thoracic surgeons would receive their specialist training as postgraduates. They should serve at least two years as surgical first assistants or registrars and in addition devote another two years to the study of surgical chest diseases. Some of this time they may want to spend abroad, the rest in a special chest hospital or chest unit in Britain.

Chest surgery required regional organization, the memorandum argued, and an effective scheme would have to be worked out in conjunction with tuberculosis services and in cooperation with local authorities. Surgeons and nurses should devote 100 percent of their time to chest surgery. Considering the staff and equipment needed, it was unrealistic for smaller hospitals or sanatoria to provide adequate service. 'Nothing could be more productive of bad results', the authors of the memorandum wrote, 'than the occasional performance of major operations once or twice a year in each of many scattered hospitals or sanatoria'.[85] Some concentration of thoracic surgical work was inevitable; each region should have one primary thoracic surgery centre that handled all types of thoracic surgical disease; tuberculosis should not be separated from other chest diseases. The centres needed access to X-ray facilities, pathological laboratories and laboratories for research; in university towns they should be part of or closely affiliated with teaching hospitals. Centres located in teaching hospitals should either be primary regional units with 50 to 100 beds or smaller units serving the requirements of undergraduate teaching. Even if the primary unit was located elsewhere, it should have a close affiliation with a university. Most regions should have one or two secondary or branch units, one of which may be located in a larger sanatorium, where the necessary staff and equipment were available.

The reality looked somewhat different than this scheme. The authors of the memorandum deplored that in many centres in the country there was no provision for even the simplest methods of treatment by collapse therapy and many institutions did not own an X-ray apparatus or were able to provide artificial pneumothorax treatment.[86] And there were not enough specialist beds available. Prior to the war, the two chief chest hospitals in London, for example, had places for 150 surgical patients. The EMS centres provided an additional 450 beds, and still there were long waiting lists. After the war, some of the EMS centres would have to be returned to their original purpose. In the early 1950s, after the launch of the NHS, in the eyes of the Society of Thoracic Surgeons the provision of thoracic surgery services was still uneven and far from satisfactory.[87] While in the North West Metropolitan area, for example, there were 8½ consultant posts for a population of 3,848,000 (0.45 million per consultant), in the North East Metropolitan area three consultants were in charge of 3,002,700 people (one million per consultant). Similarly in the regions: in Newcastle, Leeds, Sheffield, East Anglia and Liverpool, each consultant covered a population of between 0.45 and 0.6 million. In Manchester there was only one consultant per 1.1 million, in South Wales one per 1.05 million and one per 1.4 million people in Oxford. The Society recommended the establishment of 38 new consultant posts (33 in England and Wales). These were the figures. In the following section I will take a closer look at the realities facing chest services in the provinces.

Provincial realities

The Liverpool service under Hugh Reid and later Morriston Davies was a good example of a chest surgery service outside the metropolis, with Morriston Davies's unit housed in the former municipal tuberculosis sanatorium at Broadgreen. Other examples were the Manchester service, which I will discuss below, or services that other founding members of the Society of Thoracic Surgeons established in other provincial cities, for instance Mason in Newcastle, Allen in Nottingham or Armitage in Leeds. These services were mostly housed in local chest hospitals, by surgeons who also held (usually honorary) appointments at teaching hospitals. Local arrangements differed as they were results of negotiations between the surgeons and local authorities. Most provincial chest hospitals were not charity hospitals like the Brompton, but municipal institutions, often with origins in the poor law system. Quite common were tuberculosis sanatoria turned into hospitals. The main chest hospital in Manchester, for example, since the interwar period was Baguley Hospital, the precursor of today's Wythenshawe Hospital on the southern edge of the city. Patients in and around Manchester were referred for thoracic surgery either to Baguley or to the Royal Infirmary on Oxford Road. Baguley became the official regional centre for thoracic surgery in the late 1930s, with beds reserved for patients from the surrounding local authorities.[88]

Baguley Hospital had opened its doors in 1902 as a 100-bed municipal hospital for infectious diseases, run by Withington District Council. When Withington in 1904 was incorporated into the city of Manchester, the hospital became one of Manchester's municipal hospitals. Following the 1911 Insurance Act, which provided funds for the treatment of tuberculous patients, the fever hospital was converted into a municipal tuberculosis sanatorium. Baguley remained a sanatorium, with by then well over 300 beds, until the Second World War. In 1939, still before the outbreak of war, the Manchester Corporation began to build an emergency hospital – rows of temporary huts, some wooden, some brick structures – in the grounds of the sanatorium. This was part of the Emergency Medical Services (EMS) provision, an enormous planning exercise initiated in the late 1930s, when fears were growing about the imminence of war and the potentially devastating numbers of air raid casualties that were to be expected. At Baguley (as in other sanatoria in the country), with the outbreak of war, civilian tuberculosis patients were moved out to make space for war casualties. Thoracic surgeons were hoping that such extra capacity created within the EMS would be maintained after the end of the war and used to set up specialist services.

In the end, while cities like Manchester and Liverpool suffered some bad air raids, the expected deluge of air raid casualties did not materialize. The Emergency Hospital closed in 1945. Many of the temporary wards remained empty and Baguley Sanatorium returned to its original purpose, the provision of care to tuberculosis patients. Some of the more than 500 beds were under the care of a Resident Surgical Officer and reserved for surgical cases from neighbouring County Boroughs under the new Co-ordinated Thoracic Surgery scheme. Under the direction of the surgeon, the founder and long time secretary of the Society of Thoracic Surgeons, Alexander Graham Bryce, Baguley had become a regional centre in this scheme, despite conditions for surgery that were far from optimal: the theatre block was far away from the wards, in the converted former night staff resting rooms, and the lighting in the theatres was so inadequate that surgeons had to resort to flexible lights hooked into the chest cavity and battery-powered head lamps.[89] While the majority of Baguley patients in the late 1940s still received sanatorium treatment for tuberculosis, an increasing number of the cases referred to the surgeons were lung cancer patients.[90] John Dark, who joined the Baguley staff as a Medical Officer to the Surgeons (the equivalent of a registrar) in 1948, remembers that the work of the surgeons consisted of about 60 percent tuberculosis and 40 percent non-tuberculous diseases, 'of which lung cancer formed a fair complement', along with bronchiectasis in children and young adults.[91] Not only were the working conditions in Baguley poor, as we have heard, there was also a shortage of thoracic surgeons serving the Manchester area. The only consultants in this region of four and a half million people were Graham Bryce and Frank Nicholson, and they did not have full-time appointments.[92]

In 1948, with the launch of the NHS, the former Emergency Hospital was taken over by the South Manchester Hospital Management Committee, and plans from the late 1930s were revived to open a general hospital on the site. In 1952, Wythenshawe Hospital was established in the old EMS pavilions (which despite new flower beds maintained the charm of an army camp), to serve the new garden city of Wythenshawe, which provided council-owned housing for many families relocated in the course of 'slum clearances' in the inner city. The 200 patients cared for at Wythenshawe included 75 from the Christie Cancer Hospital. The neighbouring Baguley Hospital remained a tuberculosis sanatorium. A new chest clinic was opened in 1952, with the Lord Mayor wondering on the occasion 'how many people in Wythenshawe have chest trouble as a result of living in the city's crowded areas for so long'.[93]

A new generation of cancer specialists

While British chest surgeons were thinking about reorganizing chest services, a new generation of young physicians and surgeons decided to become cancer specialists. They were to shape cancer medicine in Britain (and to some degree clinical research) from the late 1940s roughly to the 1970s, when they retired. One member of this generation was David Smithers, the co-founder of a joint clinic for lung cancer patients which I will discuss in detail in Chapter 5, run by Smithers at the Royal Cancer Hospital jointly with chest physicians at the Brompton Hospital. In his memoirs Smithers positions himself in the clinical science tradition established by Thomas Lewis at the MRC-funded research unit at University College Hospital.[94] In an interesting article on 'Clinical Cancer Research' in 1956, Smithers characterizes his position as that of a middle-man between fundamental and clinical research.[95] Inspired by Lewis, Smithers had originally wanted to go into cardiology, but while working as an assistant in radiology at the Royal Cancer Hospital he became involved with the X-ray treatment of cancer patients, a branch of the work in radiology to which his boss, a professor of diagnostic radiology, did not pay much attention.[96]

When Smithers started at the Royal Cancer Hospital in 1936, he remembers, he found the hospital controlled by surgeons: 'The complete dominance of hospital practice by a small band of occasionally visiting surgeons whose word was law', he commented in 1989, 'is hard to imagine in the context of today's coordinated cancer service'.[97] Most of their professional lives, according to Smithers, were spent in undergraduate teaching hospitals and private practice. There was no radiotherapy department at the hospital, just the diagnostic radiology unit in which Smithers worked, a small radium unit that handled the radium seeds used by surgeons, and a teleradium apparatus that was run as part of a research project. Radiotherapists had no beds which they controlled, and no outpatient clinic of their own; they followed the instructions of surgeons and only saw patients again if the surgeons arranged this. Smithers found that 'the records were poor; there was no comprehensive disease index; no adequate follow-up system, and no organised social service'.[98] There was an excellent pathology department (which was to turn into the Chester Beattie Institute under Ernest Kennaway), but the research undertaken there did not have much impact on the effectiveness of cancer care.

Smithers compared what he experienced at the Royal Cancer Hospital to what Ralston Paterson, for example, had achieved at the Christie Hospital in Manchester and was particularly impressed by the precision of what came to be known as the Manchester dosage system. He also studied the approaches, imported from Paris, which Brian Windeyer had established at the Middlesex Hospital.[99] He realized that he 'became committed to a branch of medicine [he] had had no intention of entering'.[100] When in 1941 the head of the separate radium research unit, Constance Wood, after longstanding difficulties with some of the surgeons, left the Royal Cancer Hospital to run the MRC Unit at Hammersmith Hospital, Smithers and a colleague took charge of radium therapy – initially part-time.[101] A few years in private practice in Harley Street were 'a revelation'. The income was 'fair', but Smithers 'found the life disagreeable'.[102] Like other members of his generation who were attracted to clinical research, such as the respiratory physician J. G. Scadding or the clinicians and cardiovascular specialists George Pickering and John McMichael, he was not keen on private practice. He came to the conclusion that modern cancer treatment and private practice did not go well together; cancer therapy needed a well-equipped and well-organized hospital environment, with close collaboration between specialists.

By the early 1940s change was in the air, as 'a planned coordinated attack based on more accurate diagnosis and classification, better assessment of the degree and directions of spread of disease, critical reviews of past success and failure, and trials of new methods and of varying combinations of treatment were developing'.[103] The models were the centres in Paris and Stockholm or the Mayo Clinic, along with the Christie Hospital in Manchester or the Middlesex.[104] Smithers wrote a book on the radiation treatment of accessible cancer and compiled a volume on the new punch card index he had helped establish at the hospital, together with the surgical registrar Katherine Branson and the statistician Herman Otto Hartley.[105] While on call during the London bombings, he also devised a plan for the future development of the hospital and, in 1943, produced a memorandum for the reorganization of its services. Smithers recommended the establishment of specialist units, giving all visiting consultants beds and the right to run cancer clinics in their specialties; the study of brain tumours in conjunction with the Atkinson Morley Hospital; the establishment of a thoracic unit with the Brompton Hospital and a gynaecological unit with the Chelsea Hospital for Women; the amalgamation of the radium unit and the X-ray treatment into one radiotherapy department, along with full consultant status for radiotherapists; a separate diagnostic radiology department; the reorganization of the medical records, card index, and follow-up systems; and the establishment of a social service for cancer patients.[106] All but one of his recommendations, as he reports proudly in his memoirs, were eventually adopted, not least thanks to support by the surgeon and radiotherapist Stanford Cade, another pioneer of the team approach to cancer therapy.[107] Smithers was appointed to the London University Chair of Radiotherapy at the Institute of Cancer Research and became the director of the new radiotherapy department at the Royal Cancer Hospital.[108] We will hear more about his work at the Brompton and Royal Cancer Hospital lung cancer joint clinic in Chapter 5.

Conclusion: The most brilliant film on surgical technique

To conclude this chapter, it may be worth reflecting on the state of the art in the treatment of lung cancer in the 1940s. A remarkable document, a film on thoracic surgery depicting the (idealized) pathway of a lung cancer patient at the Brompton Hospital provides me with a unique window to what appeared to be an already fairly standardized set of practices, only a decade after Graham's first successful pneumonectomy. The British Council produced this documentary in 1943 to sing the praises of British medicine abroad.[109] The film had a rather sober title, 'Surgery in Chest Disease'[110] and was the first of this kind produced by the Council. It depicted the diagnostic and therapeutic journey of a lung cancer patient, starting with a mass radiography screening at the workplace. This relatively cheap radiographic method for detecting lung lesions had been developed in the 1930s by a Brazilian doctor, Manoel de Abreu, in collaboration with the Siemens electrical company.[111] The film followed the patient through hospital procedures culminating in the removal of a lung (performed by the Brompton's consultant surgeon, Tudor Edwards), showing convalescence and rehabilitation in a hospital in the countryside, and ending with the surgeon telling the patient that he was fit to return to work. The responses to initial showings of the film to selected audiences in the summer and autumn of 1943, both published and unpublished, were enthusiastic. An internal memo called it 'a picture which is almost certainly destined to become a text-book on the subject and may, with the passing of time, be regarded as a classic'.[112] Arthur Elton of the Ministry of Information congratulated A. F Primrose, the Secretary of the Council's Film Department 'on the most brilliant film on surgical technique that I have ever seen'.[113] Primrose agreed, commenting that: 'Personally I find that this film grips my attention far more than many so called thrillers'.[114] The review in the *BMJ* describes the diagnosis and treatment of lung cancer as shown in the film as a model of medical rationalization, making the Brompton Hospital appear like a well-oiled medico-surgical machine:

The main case depicted in 'Surgery in Chest Disease' illustrates a striking advance in surgical practice. Ten years ago a diagnosis of cancer or the lung was a death warrant from which there was no escape. To-day suitable cases can be cured and restored to full working capacity by surgery, though the operative risk is still great. Further, the film as a whole illustrates the growing importance of special techniques and of team work in modern medicine. Surgeons, physicians, anaesthetists, radiologists, pathologists, resident medical staff, sisters and nurses, physiotherapists and hospital almoners – all play an essential part in a complex series of processes which result in the saving of a life. The actual operation takes its logical place among a series of other special measures which precede and follow it, and which are essential to its success.[115]

For the majority of patients the reality looked very different from the ideal depicted in the film, as I will discuss in Chapter 5. Treatment outcomes remained dire, if patients received any treatment at all. Nevertheless, if there ever was a moment of hope in the history of lung cancer, this was it. Lung cancer was increasingly central to the work of chest physicians and thoracic surgeons, not only in Britain. In New York City, for example, the Memorial Cancer Hospital established a Thoracic Surgical Service in 1940, acknowledging the increasing prominence of lung cancer among the cases referred to the hospital. The annual report cites as factors leading to the establishment of the new service a resurgence of interest in the surgical treatment of lung cancer (following reports on successful operations such as the pneumonectomy performed by Graham), much increased patient safety and lowered operative mortality due to the use of sulphonamide drugs, and new anaesthetic techniques and technologies.[116] Also at Memorial and its Sloan Kettering Research Institute, the pioneers of cancer chemotherapy, David Karnofsky and Joseph Burchenal treated advanced lung cancer patients found to be inoperable and not likely to benefit from radiotherapy with injections of a cytotoxic substance, nitrogen mustard.[117] This highly experimental new treatment did not prolong patients' lives but provided some of the foundations for the development of cancer chemotherapy treatment regimes.[118] In Britain, cancer chemotherapy remained somewhat marginal until the late 1960s.

Like their colleagues in Britain, chest surgeons at the Memorial Cancer Hospital and elsewhere in America experienced rapidly increasing workloads due to an increase in numbers of cases applying to their hospitals for care. Like their British colleagues, they came to the conclusion that this increase was not a local anomaly but signalled an alarming increase in the incidence of lung cancer. By the 1940s, thus, new surgical techniques along with radiotherapy and experiments with chemotherapy had given rise to hopes that lung cancer may be treatable, while at the same time ever greater numbers of patients increased the visibility of the condition. The debate about this increase and the search for explanations are the subject of the next chapter.

4
Science, Medicine and Politics: Lung Cancer and Smoking, circa 1945 to 1965

By the 1940s, as I have argued in the previous chapter, surgery of the lungs was turning into routine, not only for tuberculosis but also for lung cancer. Meanwhile there were signs that the status of lung cancer was changing radically. Lung cancer was about to turn from a chest disease, whose significance in public health terms was relatively marginal, into a major cancer, and the only one whose main cause was known. The historian Matthew Hilton writes in his book on *Smoking in British Popular Culture* that 'the damage done to health through smoking has increasingly come to dominate the meaning of tobacco'.[1] It is important for the argument of this book that, in turn, the meaning of lung cancer has also come to be dominated by its link with smoking. This was a turning point, making lung cancer essentially different from other malignant diseases.

The process that led to today's more or less universal recognition of the association between smoking and lung cancer was long and drawn-out. It involved the application of new epidemiological methods and statistical tools suitable for research on chronic disease, and of a new concept, that of the risk factor.[2] Cigarettes did not kill every smoker, and when confronted with the bad news, there was always somebody to whom people could point, who had smoked heavily and lived until 70, 80 or 90. Not everybody indulging in the habit developed lung cancer, and for those who did, the process took decades. This long latency period made it difficult to convince people of the link between tobacco and lung cancer. By the 1950s, as we will see in this chapter, there was good evidence that this link existed, evidence that in other contexts would have been considered conclusive. But smoking was a habit in which up to 80 percent of British men indulged (including nearly 90 percent of male doctors over 35[3]), which was served by a powerful industry, and which provided significant contributions to government coffers by way of a tax that was easily collected.[4] To the Treasury, the income from this source amounted to a good 600 to 650 million pounds per annum in the 1950s (more than 15 percent of central government revenue). The economist Harvey Cole estimated that by 1963 the receipts out of tobacco duties – by then more than 900 million pounds – were enough to meet the combined Central Government expenditure on roads and health, or on education.[5] 'This dependence of the Government on revenue from tobacco', he suggested, 'is the central economic fact about the commodity'.[6]

Before the mid 1940s, only few experts noticed an increase in lung cancer incidence. After the end of the Second World War, it was increasingly difficult to ignore. In this chapter I will introduce the debate prompted by the first reports about a rise in lung cancer cases and the evidence emerging that it could be caused by smoking cigarettes. I will then discuss the broader political and cultural developments around smoking and health in the 1950s and 1960s, and the consequences for the status of lung cancer. The main focus will be on lung cancer and Britain, so it would go beyond the scope of the chapter to cover the nuanced, sometimes acrimonious debates among historians over smoking, health, and the role of history in the United States.

Observations and suspicions

In the 1920s, some surgeons, physicians and pathologists observed that cases of lung cancer seemed to appear with increasing frequency. In 1923 the German Pathological Society discussed the question at their annual meeting in Göttingen, and in the following years a number of papers dealing with the possible increase in lung cancer in different locations were published. The context was one of increasing interest in the pathology of cancer across Europe and North America, associated both with the new treatment modality of radiotherapy and research on the causes of the disease.[7] Research specifically on the incidence of lung cancer included studies by John Bright Duguid on Manchester and, prompted by Duguid's paper, Georgiana Bonser on the city across the Pennines, Leeds.[8] Duguid, then a Lecturer in Morbid Anatomy at the University of Manchester, set out to test if lung cancer was particularly common in the city and increasing, with an expected link to air pollution. He found that figures from autopsy registers going back to the 1880s and case incidence in the wards of the Manchester Royal Infirmary in the 1920s did not lead him to conclusive results. The incidence of lung cancer detected at the Manchester Royal Infirmary (as a percentage of all post-mortem examinations) had risen from 1.58 to 2.57 since the 1880s, but Duguid acknowledged that there was significant scope for error, admitting that 'the pathological conditions which are most common in man are not those most commonly revealed in the post-mortem room'.[9] Bonser found that in Leeds the incidence was slightly lower than in Manchester and there had been no increase in intrathoracic cancer at post-mortem since the 1890s. It was more common in men than in women (3.5 to 1), she observed, and there was no relation between the occupation of the patients and the disease. A *Lancet* editorial commenting on the article by Duguid in the same issue conceded that 'there is certainly an unpublished impression among the pathologists of several other British hospitals that they have observed a similar increase'.[10] However, 'when an attempt is made to translate this impression into the neat statistics which are so desirable a good many difficulties arise, and this perhaps explains why more has not been published on what seems to be an extremely interesting and important subject for cancer research to engage with'.[11]

In two articles in 1932 and 1935, A. E. Sitsen, formerly director of a Dutch-Indian Medical School in Java and then at the Institute for Pathological Anatomy of the University of Innsbruck, Austria, doubted the reality of the increase in a review of some of the international pathological literature.[12] Too many factors were variables. To start with, the demographic composition of the population had changed quite fundamentally over previous decades, and it was known by then that lung cancer affected especially older men. More people lived to old age due to a general improvement of life conditions, and the First World War had decimated the number of young men, leaving a higher ratio in the susceptible age group. But even if one disregarded such changes in the general population, claims based on post-mortem findings were not reliable, as these figures were also susceptible to changes in the composition of hospital populations. What influence, for example, did the introduction of sickness insurance have? Not every hospital patient died in hospital, and not all of those who died in hospital were subject to post-mortem examinations. In some places, some wards were more likely to send deceased patients to the pathologist for an autopsy than others. In short, even if some pathologists found more lung cancer cases in autopsies, this observation did not allow reliable conclusions about the incidence of the disease.

The director of the Research Institute of the Royal Cancer Hospital and pioneer of research on carcinogenic chemicals, Ernest Kennaway and his wife Nina, drew on death certificate data for an article they published in 1936 on the increase in the incidence of cancer of the lung and larynx between 1921 and 1932.[13] They, too, were cautious about viewing the phenomenon as real and concluded that the increase in recorded cases 'may be due to (1) an actual increase; (2) improvement in diagnosis; (3) fashion in diagnosis; or to any combination of these factors'.[14] The Kennaways compared lung cancer with prostate cancer, for which a similar increase had been recorded over the same period of time, and for which studies had shown that malignant tissue could be found in a high proportion of prostates if pathologists knew what to look for and looked accordingly closely. Was it possible that the same had happened in the case of lung cancer? However, the incidence curve for prostate cancer appeared to level off, while the figures for lung cancer continued to rise. The Kennaways wrote that 'it will be of great interest to see the changes in the prevalence of cancer of the lung in, say, the next 10 years'.[15] They were right: lung cancer was about to turn from an interesting conundrum for cancer researchers into a major public health issue.

Coughing and wheezing in the 1950s

A retrospective glance at the cancer mortality statistics for the twentieth century explains why it was easy to be unsure about the reality of the rising incidence of lung cancer in the interwar period, but increasingly difficult after the end of the Second World War. While the number of deaths attributed to the disease was rising exponentially among men from circa 1920, the standardized annual death rates were still comparably low, below ten deaths per annum, per 100,000 population prior to 1930, compared to almost 300 attributed to cancers in other locations or 150 from tuberculosis. In public health terms, lung cancer was relatively insignificant. Towards the 1940s it became more difficult to ignore. Still, in the 1950s, mortality from colorectal and stomach cancer was far higher than from lung cancer.[16] In a study concerning patients with cancer nursed at home, researchers for the Joint National Cancer Survey Committee of the Marie Curie Memorial and the Queen's Institute of District Nursing found in 1952 that 30 percent of the patients they surveyed – predominantly women as men tended to be cared for not by the district nurse but by their wives – had sought medical advice for abdominal pain, vomiting, constipation or diarrhoea, symptoms pointing to problems with the digestive tract, compared to 2 percent who consulted the doctor about a cough.[17] By this

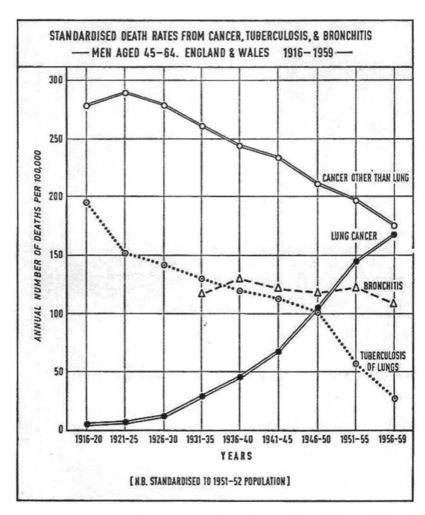

Figure 4.1   Death rates from cancer, tuberculosis and bronchitis, 1916–1959. Reproduced with permission.[18]

time, however, in the main risk group: men in middle age, around 1950, the number of deaths attributed to lung cancer had for the first time exceeded the number of deaths from tuberculosis (see Figure 4.1).

The situation further complicated by the fact that people coughed and wheezed due to many other causes, making it seem almost normal to suffer from respiratory complaints, and possibly masking some of the increase in lung cancer (see Figure 4.2). The early symptoms of

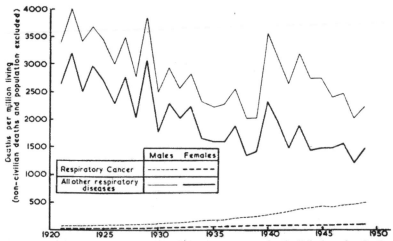

Crude annual death rates for cancer of respiratory system compared with similar rates for all other respiratory diseases (1921–49). (Taken from the Registrar-General's Statistical Reviews of England and Wales.) Since 1940, when multiple causes were certified, deaths have been assigned to the cause selected by the certifying doctor instead of by arbitrary rule.

Figure 4.2 Death rates for cancer of the respiratory system compared with rates for all other respiratory diseases, illustrating why lung cancer remained largely invisible until well into the 1940s. Reproduced with permission.[19]

lung cancer were difficult to distinguish from those of other, less fatal and much more common respiratory problems.[20] With tuberculosis incidence continuing its steep decline, what were the most prevalent respiratory problems in the 1950s?

In the 1950s the respiratory tract accounted for a considerable portion of the work of General Practitioners. John Fry estimated in 1955 that in his South East London practice 25 percent of the total work was dealing with respiratory illness.[21] Allen Daley, the Medical Officer of Health of the London County Council included respiratory infections among the 'most important diseases from the national point of view'.[22] In industrial centres in the US the situation was similar.[23] Among these respiratory illnesses, the worst and most prevalent, besides the common cold, were acute and chronic bronchitis. Curiously, there was no generally accepted definition of chronic bronchitis. It usually involved cough, sputum and some degree of disability, and was the default diagnosis where no other cause could be identified for persistent respiratory symptoms.[24] Fry reported that chronic bronchitis accounted for about 8 percent of the attendances in his practice. It was also the commonest cause of sickness incapacity in men reported to the British Ministry of Pensions and National Insurance in the fiscal year 1953–1954.[25] The author of a *British Medical Journal* editorial in November 1959, entitled 'Another Winter of Bronchitis', estimated that 15,000 to 20,000, predominantly men, were likely to succumb to respiratory problems in the following six months.[26]

Epidemiologists found that in the industrial towns and cities of North-west England the problem was even more common than it appeared to be in Fry's London practice. A study undertaken by Ian Higgins and colleagues at the Medical Research Council (MRC) Pneumoconiosis Research Unit on a random sample of the male population aged 55 to 64 in Leigh, Lancashire, revealed that only a third of all men (excluding miners and ex-miners) were free from chest problems.[27] Thirty percent reported cough, 33 percent sputum, and 18 percent both cough and sputum. Thirty-eight percent were wheezing, 36 percent reported tightness of the chest, and 23 percent both wheezing and tightness. Seven percent felt breathless. For miners and ex-miners the situation was considerably worse; 59 percent of them were wheezing, 53 percent felt tightness, 43 percent reported both, and 25 percent felt breathless. There was much wheezing and coughing in 1950s Britain, and only a small proportion of it was due to lung cancer.

John Fry, the South East London GP, suspected atmospheric pollution to be responsible for the great prevalence of chronic bronchitis and other respiratory problems, and his suspicion was supported by the observation that 4,500 deaths were directly attributed to 'the Big Smoke', the smog that brought London to a standstill for five days in December 1952.[28] The 1952 London smog triggered a good deal of interest in atmospheric pollution and its effects on health. Lung cancer among non-smokers, in fact, appeared to be more common in cities than in rural areas.[29] American cities like Los Angeles or New York also experienced smog crises.[30] Epidemiological studies on miners, foundry workers and men in other occupations, of which a whole series were undertaken in 1950s Britain, appeared to support the hypothesis that dust and other pollutants in the air were the main culprits.[31] However, Ian Higgins and his colleagues found that in spite of the great differences in atmospheric pollution, men in the rural Vale of Glamorgan did not breathe more freely than those in the smoky Leigh, Lancashire.[32] In many cases, the atmospheric pollutants appear to have been inhaled voluntarily: the respiratory problems that were so prevalent among men in the 1950s, persistent cough and sputum, as well as wheezing and breathlessness appeared to be clearly related to smoking habits.[33] But Higgins and his colleagues reported these results at a time, in the late 1950s, when the link between smoking and lung cancer was agreed on by most experts. Ten years earlier, indeed, in the 1940s, far from being an established fact, even the increase in lung cancer incidence was still widely disputed.

One expert, who had noticed the increase in deaths attributed to lung cancer was Percy Stocks, a pioneer of medical statistics and chronic disease epidemiology.[34] Like the Kennaways, Stocks had noticed the increase in the 1930s. Stocks was then chief medical statistician to the General Register Office (GRO), appointed in 1933 after training in medicine at Cambridge and Manchester and working with Pearson as a medical statistician at the Galton Laboratory at University College London (UCL). His new position at the GRO gave him direct access to death certificate data, and in the 1930s he compiled a series of annual reports for the British Empire Cancer Campaign (BECC) on the distribution of cancer of various organs in England and Wales. In 1921 a total of 186 deaths among women and 361 among men had been attributed to lung cancer. In 1945, death certificates recorded 1,480 female deaths and 5,982 male deaths from lung cancer.[35] Death certification, however, remained unreliable: as the radiotherapist David Smithers pointed out, only 11 percent out of 3,280 cases of lung cancer registered from 46 centres in England and Wales in 1945 and 1946 had had a post-mortem confirmation of the diagnosis, and in only 38 percent the diagnosis was confirmed histologically.[36] The Kennaways, in a follow-up publication to their 1936 paper, were less circumspect than they had been.[37] While they still allowed for the theoretical possibility of an artefact, they used a small study on the statistical effect of diagnostic error in stomach cancer communicated to them by a colleague, to suggest that it was unlikely that this kind of increase could be attributed to changes in diagnostic practice or false diagnoses.

In 1946, Stocks, after consulting with the surgeon, gynaecologist and radiotherapist Malcolm Donaldson, Chairman of the Statistical Committee of the Radium Commission and of the Clinical Research Committee of the BECC, convinced his superior at the GRO to write to the Ministry of Health proposing a study to investigate if increasingly frequent X-ray examinations in the context of mass radiography programmes had something to do with the increase in lung cancer deaths.[38] The Ministry forwarded the suggestion to the Medical Research Council (MRC). The MRC Secretary, Edward Mellanby was in favour of a broader statistical investigation on cancer of the lung, although he doubted that this would throw much light on its aetiology: 'It might be interesting to know, not only the relative incidence of cancer of the lung in different trades, but also whether there is any difference between town and country dwellers, and whether smoking, especially cigarette smoking is of importance, etc.'[39] Maybe the MRC could hold a conference. Could Stocks get in touch with Austin Bradford Hill about this? Hill was then the Director of the MRC Statistical Research Unit at the London School of Hygiene and Tropical Medicine (LSHTM) and statistician extraordinaire of the MRC.[40] When Frank Green, the Principal Medical Officer of the MRC contacted Hill about the plans, Hill suggested that mass radiography examinations 'are not the sole answer though no doubt contributory'. Men were much more greatly affected than women by the increase in lung cancer deaths, while both genders were equally subjected to mass radiography screening. He also thought an informal conference was a good idea.[41] Stocks prepared a ten-page conference brief, in which he concluded that 'either smokiness or pollution of the atmosphere in certain towns is an important causative factor for lung cancer or else that sunshine is an important preventive factor'.[42] Cities and towns, especially in the north, Stocks had observed, were more affected than rural areas, especially in the south.

Statistics and cigarettes: The quest for a cause

A small, informal MRC conference on lung cancer was held on 6 February 1947 at the MRC Headquarters in Old Queen Street, Westminster (for a list of participants, see Table 4.1). The conference was a crucial event in the history of lung cancer in Britain: it marked official recognition of the alarming increase in deaths attributed to the disease, which appeared to be worse than elsewhere, and the feeling that it was necessary to do something about it. It pointed to the likely involvement of cigarette smoke, even though it was still assumed that some of the increase was due to better diagnosis of primary tumours. Kennaway was not convinced by Stocks's hypothesis involving atmospheric pollution and sunshine, as pollution was decreasing and smoke in the air could hardly be accountable for the observed increase in lung cancer rates in Switzerland. He suggested that cigarette smoke might be responsible. The conference decided that 'certain members', including Kennaway, Bradford Hill and Stocks should meet and plan a 'large-scale statistical study of the past smoking habits of those with cancer of the lung', a case control study with two control groups, patients suffering from other cancers and diabetes.[43] Subsequently they decided that, rather than diabetes patients, they should interview a random selection of patients in the right age groups. Richard Doll later remembered that he initially assumed that other factors (including air pollution) were accountable for the increase in lung cancer, but a memo outlining the proposed investigation by Hill, Kennaway and Stocks states clearly that 'the main problem at issue is the possible association between the incidence of cancer of the lung and tobacco smoking'.[44] It was assumed that Kennaway was going to focus on those aspects of the investigation that could be done in the laboratory, for example on the effects of various chemicals in cigarette smoke, while Bradford Hill and Stocks concentrated on the statistical study, for which Hill requested the help of two social workers and a physician trained in statistics. In summer 1947 Mellanby briefly considered transferring the study to the new Social Medicine Research Unit at the Central Middlesex Hospital, but Hill convinced him to leave it with his unit at the LSHTM.[45] He suggested employing the young Richard Doll, who at the time worked as a research assistant to Francis Avery Jones at the Middlesex Hospital on an MRC funded study on occupational causes of peptic ulcer. Doll had attended Hills' course on medical statistics at the LSHTM and Hill was a member of the committee supervising the ulcer study.

A second conference was held on 29 September 1947, where details of the planned statistical study were discussed.[46] Kennaway received an MRC grant to purchase a spectrometer and pay for assistance, to investigate atmospheric pollution as a possible environmental factor in the causation of lung cancer, to which all sections of the population are exposed in the same way, which would help explain the absence of class differences. They examined samples of suspended matter in known volumes of air collected in Bilston, Bristol, Burnley, Kingston upon Hull, Liverpool, Manchester, Leicester, Sheffield and London.[47] Stocks also continued to investigate the influence of air pollution and the apparent association between urban residence and lung cancer incidence. Doll started his work with Hill on their statistical study in early 1948. A questionnaire was devised, which asked for smoking history, the places of residence (close to gas works?), occupational history and dietary habits (vegetarian?). A letter was drafted to hospitals in London, which asked hospitals to notify the MRC of cases of lung cancer as well as cancer of the stomach, the duodenum, the rectum and the colon. Such patients and an equal number of randomly selected control patients in the same age groups were then visited and interviewed by one of the social workers working on the study. All 14 hospitals contacted in the first round of letters agreed to cooperate. By September 1948, Doll was able to compile a confidential interim report,

Table 4.1   **Participants in the informal MRC lung cancer conference held on 6 February 1947**[48]

Conference Participants	
Percy Stocks (Chairman)	Physician and medical statistician, General Registrar's Office
Basil W. S. Mackenzie, second Baron Amulree	Physician, working for the Ministry of Health, concerned with cancer services; later known as a leading advocate of geriatric medicine in the UK
Sir Ernest Rock Carling	Chairman of the Standing Advisory Committee on Cancer and Radiotherapy of the Ministry of Health. Surgeon and developer of radiotherapy; after his retirement also member of the Radium Commission and the Atomic Energy Commission and Chairman of the International Commission on Radiological Protection
Mr Malcolm Donaldson	Gynaecologist and radiotherapist. Chairman of the Statistical Committee of the Radium Commission and the Clinical Research Committee of the BECC
Dr Frank H. K. Green	Physician, Principal Medical Officer of the MRC
Professor Alexander Haddow	Experimental pathologist and cancer researcher, Research Institute of the Royal Cancer Hospital, from 1946 Director of the Chester Beatty Institute
Professor Austin Bradford Hill	Medical Statistician, Director of the MRC Statistical Research Unit at the London School of Hygiene and Tropical Medicine
Professor Ernest L. Kennaway	Experimental and chemical pathologist, cancer researcher, Director of the Research Institute of the Royal Cancer Hospital and Chester Beatty Institute, from 1946 Emeritus at St Bartholomew's Hospital
Mrs Nina L. Kennaway	Ernest Kennaway's wife and collaborator. According to a note by her husband 'has collected all the material on this subject, and I will undertake she will not speak'*
Dr James Maxwell	Chest physician at St Bartholomew's Hospital with long-term interest in lung cancer, working part-time for the Ministry of Health mass radiography scheme
Dr Alice Stewart	Physician and epidemiologist, first assistant to John Ryle, who was unable to attend. Department of Social and Preventive Medicine at Oxford; previously worked as assistant to Leslie Witts
Dr Martin Ware (Secretary)	Physician, MRC Headquarters

*E. L. Kennaway, 'Note', 1 February 1947, FD 1/1989, UK National Archives.

which he presented on 6 October to what was now in MRC communications informally called the Cancer of the Lung Committee. 'The results appear', he wrote, 'to show a definite association between carcinoma of the lung and smoking'.[49] He observed no other associations.

Over the next year and a half, Doll and Hill contacted more hospitals, the social workers (two more had been enrolled) interviewed more patients, the sample grew, and three further confidential interim reports were compiled. By the end of 1949 they had data from 709 pairs of lung cancer and control patients and felt they were ready to draft a journal article. When they showed the draft to Harold Himsworth (Mellanby's successor as MRC Secretary), Himsworth urged them to check if they would come to the same conclusions outside London. They contacted hospitals in and around Bristol, Cambridge, Leeds and Newcastle. Before they had collected and evaluated the new data, in spring 1950, Ernst Wynder and his sponsor Evarts Graham published their US-based case-control study in the *Journal of the American Medical Association*, about which the British researchers knew nothing.[50] Hill used his contacts to the *British Medical Journal* to ensure that their paper was published later in the same year. Doll and Hill concluded 'that smoking is a factor, and an important factor, in the production of carcinoma of the lung'.[51] Wynder and Graham were similarly cautious, stating that 'the temptation is strong to incriminate excessive smoking over a long period as at least one important factor in the striking increase of bronchiogenic carcinoma'.[52]

The US study by Wynder and Graham had somewhat different origins from the one undertaken by Doll and Hill under the aegis of the MRC, which may explain why they did not know of one another. As in Britain, in the US, surgeons and pathologists in the interwar period had noticed that they were encountering more lung cancer cases. Chest surgeons such as Alton Ochsner and Michael DeBakey in New Orleans or Richard Overholt in Boston had noticed that their lung cancer patients were usually smokers. Ochsner took this observation so seriously that he turned into an ardent anti-smoking advocate and prohibited his staff from smoking.[53] However, there was no systematic data collection and observations remained anecdotal. Graham, himself a heavy smoker, was initially sceptical about Ochsner's claims but changed his mind in light of Wynder's findings.[54]

German-born Ernst Wynder was a medical student at Washington University, St Louis, where Graham was professor of surgery and director of the chest clinic that he had established at Barnes Hospital. Wynder was interested in cancer research and developed ambitions to win an award for the best piece of research by a fourth year medical student. He first painted carcinogens on the backs of mice to study the development of skin cancers in the laboratory of Edward Cowdry at Washington University, then went to the Jackson Laboratory in Maine to study the genetics of mice cancers with Clarence Little. While doing a course in pathology he developed an interest in the possible link between smoking and lung cancer, apparently after observing the autopsy of a 42-year old lung cancer patient who had been known to smoke two packs of cigarettes per day. During a summer fellowship at New York University in 1948 he undertook interviews with some 20 lung cancer patients and controls, which indicated that there was indeed evidence of a link. Back in St Louis, Wynder obtained Graham's permission to interview lung cancer patients at Barnes Hospital, in spite of doubts that Graham and other senior colleagues, most of them smokers, had about Wynder's hypothesis. When Wynder spent several days during the next Christmas break at the office of the Manhattan Cancer Society, its director encouraged him to apply for a grant. He was successful, and a research grant of the American Cancer Society (ACS) allowed him to appoint an interviewer. In early 1949 he presented results of his research during an ACS conference. There were no questions or comments after his talk, neither encouraging nor critical. The paper he and Graham published in the following year was based on interviews with 684 lung cancer patients. Wynder did win the award he was aiming for in his fourth year as a medical student.[55]

The historian of science Robert Proctor has suggested that the link between smoking and lung cancer was not 'discovered' by Wynder and Graham or Doll and Hill, but by researchers in Nazi Germany.[56] In Germany, as in Britain and the US, pathologists had indeed noticed in the 1920s that they encountered more lung cancers than they used to. There was also a growing body of scholarship on the damage done to health by tobacco consumption. Work by the vigorous anti-tobacco advocate Fritz Lickint or the epidemiological studies by Franz Müller are worth mentioning. However, while in the 1950s Lickint published books and pamphlets on lung cancer in East and West Germany, in the 1930s he was as much, if not more interested in the effects that smoking had on the reproductive organs.[57] In a booklet on tobacco consumption and health he published in 1936, only one page out of 92 deals with lung cancer, and the evidence he cites is mostly anecdotal.[58] A case control study originally reported by Eberhard Schairer and Erich Schöniger in 1943 has recently been rediscovered and an English translation published in the *International Journal of Epidemiology*.[59] The study was financed by the institute for research on tobacco hazards that had been established in 1941 at the University of Jena with a grant from Hitler's Reich Chancellery for Karl Astel, a medical doctor, president of the university since 1939, and a high-ranking SS officer. Astel was a devoted Hitler supporter and, as Proctor points out, a 'rabid anti-tobacco activist'.[60] The paper was largely based on Schöniger's medical dissertation, produced at Astel's institute. Proctor and the epidemiologist George Davey Smith stress that, while morally tainted, the German research undertaken during the Nazi regime was methodologically good (Doll was more critical). It is undisputable, however, that it did not have the same impact as the post-war studies in the UK and the US (nor did, as I have argued, earlier studies undertaken in these countries).[61] This had much to do with the war and the fact that the journal issue in which Schairer and Schöniger published their results did not reach Britain for some time, but also with the simple truth, as I have argued, that until after 1945 lung cancer was not perceived as a major public health problem.

The studies by Doll and Hill and Wynder and Graham achieved legendary status as models for what epidemiological cancer research can achieve, and won their authors much praise and ultimately fame.[62] Their immediate impact on health policy was limited. But the stir they caused in the media and the studies that followed them over the years triggered a public debate which continues to the present day and which has irreversibly linked lung cancer and smoking.[63] But was the link causal?[64] Wynder and Graham turned to laboratory experiments on animals while Doll and Hill devised a large-scale prospective epidemiological study to provide new arguments.

In 1950 Doll and Hill, with the help of the British Medical Association, wrote to all medical doctors registered in Britain, inviting them to complete a questionnaire about, among other factors, their smoking habits. They succeeded in securing the participation of over 34,000 doctors for the study, and Doll together with a succession of colleagues followed this cohort for 50 years, until 2001, investigating relationships between mortality and morbidity in this group and their smoking habits.[65] In the US, starting in 1952, the American Cancer Society's chief statistician, Cuyler Hammond devised an even bigger cohort study. Hammond's idea was to enrol 22,000 ACS volunteers as interviewers. Each of these was asked to interview about ten men between 50 and 69 (over 200,000 were interviewed). Each year the volunteers were asked to report if their interviewees were alive or dead, and death certificates were obtained for those who had died. A preliminary report was presented in June 1954 at a meeting of the American

Medical Association; the results were announced in the *New York Times*, and the short-term impact was considerable: cigarette consumption per capita dropped by 6 percent in 1954, after having increased every year since the early 1930s by an average of 5.6 percent per year.[66]

Wynder, Graham and their St Louis colleagues, meanwhile, constructed a succession of machines that 'smoked' up to 60 cigarettes at a time and extracted the tars in acetone in condensing flasks. The tar-acetone solutions were painted onto the backs of standardized laboratory mice and changes in their skin analysed. Their research led them to conclude that smoking tobacco released carcinogens, explaining the results of the epidemiological studies. But the results of increasingly sophisticated experiments or epidemiological studies did not lead to closure in the controversy over the damage done to health by cigarettes. Sociologists of scientific knowledge suggest that this is not unusual for scientific controversies, especially those that come with strong interests attached to the respective positions.[67] The debate was kept open by the tobacco companies and their allies in cancer research, who argued that epidemiological research was insufficient as it only produced statistical associations and not experimental proof. Epidemiological and clinical studies, however, are the only ways by which effects on humans can be researched. Experimental studies on laboratory animals were frequently challenged because they had only limited validity for humans or did not represent the realities of smoking sufficiently well: it proved very difficult to persuade laboratory animals to puff cigarettes as humans do.[68] Commenting on the shortage of experimental evidence, a *BMJ* editorial in 1950 suggested that animals subjected to an atmosphere of mechanically generated tobacco smoke should be compared not to human smokers but to non-smokers in a smoking compartment.[69] By 1958, however, the editors of the same journal felt that sufficiently good experimental evidence was available.[70]

Smoking and politics

Considering the proportion of men who smoked in the 1950s, in Britain and elsewhere, it is perhaps not surprising that the debate was dominated by smokers and reformed ex-smokers, on both sides of the argument. When they evaluated the questionnaires they sent to British doctors for their prospective study in 1951, Doll and Hill found that nearly 90 percent of their male respondents over 35 were smokers; 62 percent of them smoked cigarettes, 17 percent pipe, and 21 percent both pipe and cigarettes.[71] Doll gave up smoking when he compiled the results of the first study and the link became obvious to him. Hill apparently still had a case with cigarettes on his office desk at the LSHTM for visitors, because it would have been impolite not to.[72] Most writers calling for smoking restrictions did so apologetically, frequently stressing that they were moderate smokers or ex-smokers themselves.[73] Concerns over the health of non-smokers were rare. Risk and the long-term implications of the habit entered the debate from the mid 1950s, when the results of epidemiological studies on smoking and health made it into the newspapers and started to inform a new agenda in public health and health education, emphasizing individual responsibility for one's health.[74]

One of the ex-smokers who shaped the initial response to the findings by Doll and Hill was Horace Joules, medical director of the Central Middlesex Hospital, chest specialist, founder member of the Socialist Medical Association and a champion of both social medicine and the prevention of respiratory diseases. Joules quickly kicked his 40-a-day habit and turned into a vocal opponent of cigarette smoking and the tobacco industry. As a member of the Central Health Services Council's Standing Committee on Cancer and Radiotherapy, Joules was crucial in pushing the Ministry of Health to respond to the emerging evidence pointing to a link between smoking and lung cancer.[75] Joules consistently spoke of a lung cancer 'epidemic'. He was also a frequent writer of letters to medical journals and newspapers. In a letter to the *Times* in 1956, which triggered a host of responses from other correspondents, he reminded readers of the paper that 'we led the world between 1910 and 1925 in our consumption of cigarettes as now we lead in the lung cancer figures'.[76] As he stated in a public talk in 1955, Joules felt outraged not only by the scandal of what he saw as a preventable epidemic, but also the fact that most of the victims of this epidemic died in ignorance about the cause.

> Until five years ago I was smoking up to forty cigarettes daily in complete ignorance of the harm which might result. My father smoked for fifty years in ignorance; he died at somewhat over seventy years of age of the disease – he died in ignorance but not without severe suffering. A medical brother-in-law is suffering from the same condition at the present time. It is essential that the rising generation should not take the risk in ignorance and the duty of lifting this veil of ignorance must rest squarely with the Ministry of Health, which should be asked to live up to its name and cease evading responsibility.[77]

Initially the Standing Committee on Cancer and Radiotherapy chaired by Sir Ernest Rock Carling (a committed smoker), was against advising the Minister to initiate a campaign against smoking, but Joules continued to push the issue. Following the publication of the first results of Doll and Hill's follow-up study, Hill was invited by the Committee to comment and explain the results. The Tobacco Manufacturers Standing Committee responded by submitting a response to the Doll and Hill study composed by Imperial Tobacco's statistician, G. F. Todd.[78] The Committee considered Todd's arguments but ultimately followed Doll and Hill, accepting the evidence for a causal link between cigarette smoke and lung cancer and advising the minister accordingly. The Health Minister of the Conservative government of the day, Iain Macleod, who so far had refused to answer requests in Parliament for a government statement on the issue of smoking, eventually followed the recommendations of his Advisory Committee. Macleod made an announcement to the Commons and held a press conference on 12 February 1954. The Committee's opinion, he announced, was that a relationship between smoking and lung cancer had been established. Although there was a 'strong presumption' that this relationship was causal, it was not a simple cause-effect relationship and there was so far no clear evidence as to how exactly smoking caused lung cancer. There was no proof, but there was an association. The main complicating factors were the difference in lung cancer incidence between town and country (as shown by Stocks), which suggested that air pollution may play a part, and the failure to identify a cancer-producing substance in tobacco smoke. While tar extracts had produced skin cancer in mice, this could not be considered conclusive evidence that tar was also responsible for lung cancer in humans. Also, no dramatic and sudden fall in death rates could be expected even if everybody stopped smoking immediately, as the development of lung cancer appeared to be a long-term process.[79] 'Young people', however, the Minister advised, 'should be warned of the risks apparently attendant on excessive smoking'.[80] He left it open to interpretation when a smoking habit was 'excessive'. And if people did not smoke excessively, were they safe from lung cancer?

Macleod's public statement was widely reported in the British press, by both broadsheets and tabloids. The *Times* dedicated a leader article to 'Smoking and Cancer' and both the *Manchester Guardian* and the *Daily Mirror* reported on page one.[81] The *Economist*, however, remarked that considering the annual death toll attributed to lung cancer (14,000) the press echo was tame compared to the publicity granted to a statement on poliomyelitis (750 deaths in its worst year so far): 'Some cynics may observe that the press does not want to lose its advertising revenue from the cigarette manufacturers; others, probably more justifiably, will remark on the widespread public dislike of facing the facts about cancer – especially when those facts concern so darling a sin as smoking.'[82] Concerns over the possible financial consequences of tobacco-critical articles or television programmes, did in fact influence editorial decisions, as the journalist Peter Taylor has shown.[83] Nevertheless, in the following months and years, British newspapers continued to report frequently on rising lung cancer death rates, new political developments, ministers' statements, and the latest scientific findings linking smoking to lung cancer, as well as those challenging this link.

Macleod found himself in a difficult position regarding the smoking-lung cancer link, and so did his successors as health minister. Apart from increasingly conclusive evidence pointing to the health risks associated with smoking, ministers faced pressure from the Treasury, where officials feared losing a reliable source of income, and lobbying efforts from the tobacco companies, whose umbrella organization offered the MRC £250,000 for research on lung cancer and smoking to signal their goodwill and interest in solving the controversy. But Macleod also faced a more direct, practical dilemma. It was well known to journalists that he was a chain smoker, which affected his credibility somewhat. Was his smoking excessive? According to one of his successors, Kenneth Robinson, Macleod was unsure what to do during this particular press conference: 'If he didn't smoke, there might be panic; if he did, they might not think he took it seriously. In the end he lit up.'[84] The Ministry of Health received a number of letters whose writers took offence at Macleod's chain smoking during the press conference – he apparently lit each cigarette from the last. One of the letter writers, a J. S. Reynolds from Ripley, Derbyshire, was representative of others in suggesting that 'this display was in bad taste, considering the nature of the announcement'.[85] Robinson himself, Minister of Health from 1964 to 1968, was a self confessed tobacco 'addict', but during his time in the Ministry of Health never smoked in public, because, as he confessed to Taylor, he had felt 'guilty all the time'.[86]

When talking about the 1950s and early 60s, the use of the term 'addict' is slightly anachronistic. The notion of addiction had not then entered the general debate over the risks of smoking; smoking was a 'habit'. Still, it was clear that it was not easy to stop smoking and education efforts concentrated increasingly on children and young people.

A survey in 1963 revealed that by then nearly one in five secondary school pupils, girl or boy, was smoking.[87] The *Daily Mirror* reported that some parents sent their children to school with cigarettes in their lunch boxes.[88]

The pressure on government increased with a widely publicized MRC statement on Tobacco Smoking and Cancer of the Lung in 1957, which concluded that the great increase in the death rate from lung cancer over the previous 25 years was primarily associated with cigarette smoking. In the Council's opinion, the evidence gathered by then in 19 different studies in 17 countries, pointed to a cause and effect relationship, a conclusion that was supported by the identification of several carcinogenic substances in tobacco smoke. Only a relatively small number of cases could be attributed to specific industrial hazards. A contribution of atmospheric pollution in a proportion of cases also had to be assumed, but the exact extent and number of cases had yet to be determined.[89] The government accepted the MRC conclusions, as announced to the House of Commons on 27 June 1957 by the Parliamentary Secretary to the Ministry of Health, John Vaughan-Morgan.[90] The government felt that it was now right to ensure that this latest authoritative opinion was publicized. All local health authorities were to be asked to inform the public of the MRC's conclusions. The risks should be made known, so that smokers could make up their own minds.

In 1962 the Royal College of Physicians (RCP) published its report on *Smoking and Health*.[91] On 2 March, RCP President Robert Platt (who had given up smoking in 1954) presented the case against cigarettes, in a conference room filled with a blue haze from reporters' cigarettes. 'Cancer of the lung', Sir Robert said, 'is not the kind of thing you joke about'.[92] The science editor of the *Daily Mirror* observed that during Platt's presentation the smoke in the room began to thin out: 'Many of the cigarettes were doused and few of the reporters lit up again.'[93] There was not very much new evidence in this report, which was part of the College's attempts, under Platt, to modernize and become more relevant to life in twentieth century Britain.[94] But the RCP committee, formed following an initiative of Charles Fletcher, respiratory physician at Hammersmith Hospital and former member of the MRC Pneumoconiosis Research Unit, had compiled the available evidence in a slim (25,000 words) and attractive volume with a brief, very readable summary, presenting, as the *Economist* put it, 'the prosecution's case in language accessible to the average juryman'.[95] Still, there was no confession and tobacco had not been caught red-handed. But the statistical evidence from by then at least 23 investigations in nine countries was clear, and the prospective studies had shown that smokers had to worry not only about lung cancer, but also about an increased risk of dying from cardiovascular problems. Some might say that only one in eight heavy smokers died from lung cancer, Sir Robert admitted in his presentation: 'This is true. But suppose you were offered a flight on an airline and were told that usually only one in eight of their aeroplanes crashed, you might think again.'[96] The RCP report called for higher taxes and smoking bans to alter the social acceptance of the habit. Children especially should be protected.

The RCP Report was followed in 1964 in the US by a Surgeon General's report on *Smoking and Health*, whose publication was widely covered in the American press (and to some degree in Britain). The Surgeon General's report represented a significant turning point in the complex political and scientific debates in the US over the status of the statistical evidence linking smoking and cancer, made more complicated by the influence on US politics of not only cigarette manufacturers (as in Britain) but also tobacco growers in the southern states.[97] If the RCP report presented the case of the prosecution, the Surgeon General's report was more like a verdict, an unambiguous 'guilty'.[98] The Advisory Committee that compiled the report in a smoke-filled conference room at the National Library of Medicine was constituted like a jury, closely scrutinized by both the tobacco industry and tobacco critics, and with as many smoking members as non-smokers. Richard Doll remembers that following the publication of these two reports on *Smoking and Health*, 'the idea that smoking was a major cause of lung cancer ceased to be seriously challenged, except by the tobacco industry outside the UK (where it had been quietly accepted) and by a few eccentric individuals'.[99]

How should British ministers respond to the increasingly conclusive evidence linking smoking and cancer? The government, argued an *Economist* leader writer on occasion of the official endorsement of the MRC report in 1957, 'could hardly be expected to sit on the fence until the carcinogenic factor in tobacco smoke had been identified – even though other overseas governments ... are doing just that'.[100] The strategy to rely on local councils was going to be 'regarded here as about the least it could decently do'.[101] It was going to be interesting to see how enthusiastically Bristol and Nottingham were going to address the issue, cities in which cigarette factories provided about 5 percent of the local jobs.[102] In the event councillors in Bristol, along with London, Manchester and Glasgow announced plans in March 1962 to ban smoking, as suggested by the RCP report, in places such as cinemas, theatres, dance-halls and on public transport.[103]

Smoking and lung cancer was different from previous public health issues. Where infectious diseases were concerned, not only the individual was at risk, but the community. 'Now, for the first time', the *Economist* wrote in 1957, 'the Government is taking action to counter a disease that is of direct concern only to the individual'.[104] At issue were questions of personal freedom, of treating smokers as adults, and also of precedents for the future. If a link was established between fat and heart disease, would official warnings be issued against eating butter and cream? (yes, they would, as we know today with hindsight) Where would it all end? While there had always been reasons for smoking bans in public places – nuisance to non-smokers, for example – the *Economist* felt that ministers could not be blamed for walking warily when it came to legislating for such bans. The government, the *Economist* editorial argued, 'should not allow itself to be pushed too hastily, and too far away, from the fence it has begun to climb down from so cautiously'. But smokers should not fool themselves either: there were good reasons for the government to get off that fence.

It remained difficult for governments to take action, in Britain and elsewhere. The climate after the RCP report favoured some form of action, but prohibition was out of the question. After the publication of the report, the *Daily Mirror* suspected rightly that the health minister Enoch Powell was unlikely to take drastic steps.[105] The Conservative government encouraged voluntary action by the tobacco industry. Within a month of the publication, the tobacco industry agreed to be bound by a code, policed by the industry itself, which excluded advertisements glorifying smoking. To protect young children, the industry also committed itself to no longer showing cigarette advertisements on television before nine in the evening, and the Carreras-Rothman company withdrew all their vending machines.[106] When Kenneth Robinson became Minister of Health in Harold Wilson's Labour government in 1964, as he told Peter Taylor, 'cigarettes were on his shopping list'.[107] He found that self-policing did not work; judging by advertisements since 1962, according to Taylor, 'the policeman was off-duty most of the time'.[108] Robinson wanted more; he aimed to stop all tobacco advertising and ultimately introduce a ban on smoking in public places. With help from his cabinet colleague Tony Benn and by involving the Independent Television Authority, all cigarette advertising was banned from British television from 1 August 1965.[109] When it came to his 'shock plan' (*Daily Mirror*), however, Robinson was unsuccessful.[110] There was intense debate in the Labour government on this issue. Drinking and smoking were pleasures of the masses, and the government had already alienated some of the Labour grassroots by introducing unpopular 'Breathalyzer' tests to stop drunk driving. While nobody any longer denied the health risks, the question was how far the Labour government should interfere with personal liberties. Should they become known, as Robinson's cabinet colleague Richard Crossman wrote in his diary, 'as the government which stops what the working classes really want'?[111] Robinson's proposals for legislation were dead when in 1968 Health became part of the Department of Health and Social Security, with Crossman as its first Secretary of State. In 1970 the Conservatives under Edward Heath won the General Elections, and the new Health Minister Keith Joseph favoured a return to the policy of voluntary agreements. In 1971, nevertheless, the year in which the Royal College of Physicians published its second, updated report, *Smoking and Health Now*, a (fairly mild) government health warning appeared on cigarette packages: 'Warning by H.M. Government: Smoking Can Damage Your Health'.[112]

A remarkable degree of public awareness

What effect did the succession of alarming official statements and warnings have on cigarette sales? The year of Macleod's press conference, 1954, was a fairly good year for the British tobacco industry, with sales increasing to their highest level since 1947, when a sharp rise in tobacco duties had resulted in a considerable fall in sales.[113] There seems to have been no decline following Macleod's announcement. The 1957 MRC statement and the following government warning caused, according to the *Economist*, 'some momentary tremulations in sales returns' but personal spending on tobacco products continued to increase.[114] There were, however, some more subtle changes: consumption per head continued to go up but people increasingly chose filter-tipped cigarettes, which were cheaper, contained less tobacco, and were assumed to be less damaging.[115] Cigars, too, were becoming more popular. The publication of the 1962 Royal College of Physicians Report led to a fall in tobacco sales in spring and early summer of that year; subsequently manufacturers noticed that sales were picking up again.[116] The suicide of a worried smoker in Gillingham reported by the *Daily Mirror* clearly was an exception.[117] Following the Report, the manufacturers increased their expenditure on advertising considerably.[118] However, cigarette sales to men stagnated; they still had not reached the pre-1962 level by the time the second RCP report was published in 1971. Now women were driving up sales figures.[119] Smoking habits were getting less gendered, while class differences became more visible, as I will discuss later. The advertising ban on TV in 1965 did not affect cigarette sales, which according to the *Economist* were 'booming' after this year.[120] The ban did not even make it more difficult to introduce new brands, as initially assumed by the manufacturers. The second best selling brand in the late 1960s, 'Player's No 6' was introduced after 1965. Cigarette manufacturers found other effective ways to advertise their products, notably the sponsoring of (motor) sports and other events. British smokers, it seems, were not very responsive to official warnings. But decisions by the Chancellor of the Exchequer did show effects, albeit short-lived. Increases in tobacco duty, for example in 1956, led to temporary declines in sales.[121] However, even though by 1960 tobacco duty was equivalent to a purchase tax of 425 percent, this did not appear to put a stop to growth in cigarette sales.[122]

If ministerial statements or warnings by medical organizations did not persuade smokers to reduce their cigarette consumption, were more sophisticated education campaigns needed to inform citizens about the dangers of smoking? The Edinburgh Health Department launched an extensive anti-smoking campaign in November 1958, in response to the recommendations by the Ministry of Health and the Department of Health for Scotland that local authorities should educate the public about the health hazards associated with cigarette smoking. Adverts in which the Medical Officer of Health warned of the dangers of smoking were placed in all newspapers; the three major Edinburgh papers also gave the campaign their editorial support, providing space in news and gossip columns, on the sports pages and in photographic sections. Posters were displayed on boards throughout the city and in factories, offices, clubs, clinics, consulting rooms, libraries, public baths, and shops; a poster was even shown in all Edinburgh buses. Thirty thousand copies of the BMA booklet *Smoking – the Facts* were distributed, as well as 150,000 copies of an Edinburgh campaign leaflet specially designed by 'one of Britain's foremost display artists'; and a 'clear and forceful letter from the Medical Officer of Health' was sent to every Edinburgh household.[123] Seventy public meetings were organized all over the city in cooperation with various associations and clubs. Twenty-four meetings were held for over 2,600 teachers, who were told about the case against smoking by a doctor using various visual aids and showing them the BBC film *Facts and Figures – Smoking and Lung Cancer*. Every class of secondary school children received a visit by a member of the Health Department. Meetings were also arranged with nearly all parent-teacher groups in the city, during which the film *1 in 20,000* was shown. There was also a rally of voluntary workers, a concert for young people, a Sunday cinema meeting and radio and television publicity. Ann Cartwright and F. M. Martin, who were then based at the Department of Public Health and Social Medicine at Edinburgh University, were invited to evaluate these efforts together with J. G. Thomson, Edinburgh's Medical Officer for Research and Health Education.[124] What then was the effect of this campaign?

The people of Edinburgh had certainly noticed the campaign, but Cartwright, Martin and Thomson found no indication that it had persuaded people to give up smoking. Nine percent of the smokers thought that the campaign had somewhat influenced their smoking habits, but when the researchers compared the actual levels of tobacco consumption in their sample before and after the campaign, they found no difference. The campaign may have succeeded in persuading a number of them to try reduce their smoking, but such attempts appeared to be less successful than individual smokers assumed. All but 2 percent of the informants had heard about the postulated link between smoking and lung cancer already before the beginning of the campaign, so there was not much to be gained with regard to public awareness. The campaign provided an opportunity to persuade those who were not convinced that the hypothesis was correct; the researchers found, however, that the proportion of Edinburgh citizens prepared to accept that lung cancer was more likely in smokers than in non-smokers, had not increased significantly. They found a greater readiness among smokers, though, to assume that they might be affected by more common respiratory problems in the future. Attitudes changed slightly with regard to young people smoking: more informants thought that it was wrong; and to the relaxing qualities of smoking: 11 percent rejected the notion that smoking helped people to relax, compared to 6 percent before the campaign. Young people were also more likely to have noticed the campaign. All in all, however, the change in public attitudes brought about by the campaign appeared to be 'very modest'.[125] There was 'certainly nothing to suggest that smoking was becoming socially unacceptable; but there was probably a slightly wider acceptance of the idea that one ought not encourage young people to start it'.[126]

One of the more surprising findings of the Edinburgh researchers was that perhaps people did not really need educating. Already before the campaign an impressive 98 percent of their informants had heard of the theory linking smoking and lung cancer, according to the authors 'a remarkable degree of public awareness'.[127] One-third of those who smoked at the time and 50 percent of the ex-smokers assumed that smoking had already affected their own health adversely. Younger informants were more likely to assume that smoking was bad for them than older men. A great majority expressed the view that their smoking might affect the health of other people. Another interesting observation was that 'anti-tobacco' attitudes, including readiness to welcome smoking bans in certain public places (not in pubs, though), were relatively widespread among the general population and quite common among smokers, too. More than 40 percent of the smokers stated that they would like to quit smoking if this could be achieved without effort. They cited the expense of smoking as the main reason. About one-quarter mentioned health reasons other than cancer. Only 3 percent admitted to being concerned about cancer. While half of those interviewed agreed that smoking may lead to cancer, many of the informants assumed that this applied only to what they considered 'heavy smoking'. This resonated with Macleod's 1954 warning about 'excessive' smoking or an article in the *Times* reporting on a lecture by the thoracic surgeon Sir Clement Price Thomas, who is cited as stating that 'those who smoked more than 20 cigarettes a day for more than 20 years were liable to contract cancer of the lung' and that 'there could not be more than 10 percent of heavy smokers who contracted cancer of the lung'.[128] Apparently many smokers were willing to take the risk. Clearly, there was no sign of the panic and 'cancerphobia' that some feared might result from the various announcements concerning the smoking-cancer link or from campaigns aimed at educating smokers.[129]

The authors of the Edinburgh study suggested that taboos associated with cancer played a role in the surprising refusal of smokers to consider themselves at risk of suffering from the disease. While the population were familiar with everyday respiratory illnesses and symptoms, lung cancer was comparatively rare and it was known to carry a high fatality rate: three-quarters assumed that for cancer patients death was a more likely outcome than recovery.[130] As a consequence, as Cartwright and her colleagues put it, 'theories which impose a major responsibility for its causation on the individual as distinct from forces outside his control, are more likely to provide a reaction of denial and rejection'.[131] This was possibly a form of unconscious, emotional blockage:

The widespread fear of cancer – long nurtured by a conspiracy of silence – generates a resistance that factual assertions do not easily overcome. This resistance is strengthened by the feeling of guilt which may be aroused by incrimination of a habit which the individual knows that he ought to be able to control.[132]

In her contribution to an anti-smoking book edited by Charles Fletcher and published by Penguin for a wide audience in 1963, the politician and journalist Lena Jeger suggested that for most children – who after all were the main target of the government's education campaign – the threat of cancer was too far removed from what they experienced in their daily lives to influence their choices.[133] Of course cancer was a terrible word, maybe it had happened to their grandfather and might strike them in years to come. But 'for every schoolboy who knows somebody who died of cancer, there are hundreds to say that none of the smokers they know had died of this disease'.[134] One 'pretty sixth-form girl' told Jeger that: 'You've got to die of SOMETHING. ... Why not that [i.e. lung cancer]?'[135] According to Jeger: 'She was more concerned about the possibility of going through life wheezing and coughing, though much too happy and healthy to be really bothered.'[136]

The Edinburgh findings posed a real challenge to the philosophy of health education. Very nearly all informants had heard about the smoking-cancer link, already before the campaign, but this knowledge did not influence their actions in any significant way, demonstrating, as the authors of the study put it, 'the fallacy of the Socratic assumption that underlies much health education'.[137] Smokers were much more likely than non-smokers to deny the validity of the evidence. Let's look at doctors, for example. In fact, by 1962 nearly half of the male doctors were non-smokers (up from 10 percent in 1950, as Doll and Hill had found), compared to 24 percent of all other men.[138] This indicates that the accumulating evidence linking smoking to illness did have an effect. However, some of the attitudes that Cartwright and her colleagues found in their Edinburgh sample appeared to be evident among doctors, too. They also adapted their convictions to their habits: among a sample of Oxford hospital doctors who completed a brief questionnaire in 1955, 71 percent of the non-smokers were convinced by the evidence that linked smoking and lung cancer, but only 46 percent of the cigarette smokers.[139] Most of the doctors who stopped smoking said they did so to save money, not because of the link with lung cancer. Clearly, hitting smokers in the wallet with increased tobacco duties appeared to be a more promising strategy than appealing to their reason.

However slowly, and for whatever reason, the demography of smoking did change, away from a predominantly male habit that was more or less equally distributed in different social classes. The doctors were only one group of middle class professionals (admittedly also particularly well educated on this issue), where smoking became less common, especially among younger men. Already in 1958, as Cartwright, Martin and Thomson found in their Edinburgh survey, professionals and white collar workers were less likely to smoke regularly than semi-skilled and unskilled workers. The proportion of men smoking more than 15 cigarettes daily was considerably higher among the unskilled, who were also more likely to suffer from lung cancer. Men working in unskilled jobs were likely to start smoking at an earlier age, while still in school. The heaviest child smokers, according to a survey reported in the *Daily Mirror* in 1960, were in the 'lower mental ability group' at secondary modern school.[140] Nearly a quarter of the smokers in professional and clerical occupations said that they started smoking when they served in the armed services. Non-smoking was associated with church-going and abstention from alcohol, while smokers tended to be partial to drinking and frequented pubs.[141] Not all would find non-smoking attractive. Jeger observed in 1963 that 'non-smoker' had 'acquired a slightly pejorative inflexion ..., a hint of the crank, the faddist, the neurotic, close to the vegetarian, the pacifist, fresh-air fiend, noise abater, and members of other minority pressure groups'.[142] Gender differences in smoking habits became increasingly smaller as women took to cigarettes: by the early 1960s, while the proportion of men who described themselves as smokers had declined to 75 percent, 50 percent of women were now indulging in the habit.[143] Over the decades, smoking became ever more closely associated with lower social status. By the 1970s, smoking began its transformation into a stigmatized activity, a habit that was attractive only where it marked the rebel, but otherwise associated with the uneducated and the lower classes. By 1994, 42 percent of the men and 35 percent of women among unskilled and manual workers smoked, but only 15 percent of the men and 13 percent of the women in the professional classes.[144]

Conclusion: The causes of lung cancer

The link with smoking has come to dominate the meaning of lung cancer, but we should not forget about other possible causes of lung cancer. There have always been a proportion of non-smokers affected by lung cancer. About 10 percent of lung cancer cases are thought to be unrelated to tobacco consumption. Even when the smoking-cancer link was widely accepted, there was general agreement that air pollution and exposure to dust or chemicals at the workplace were likely to play important roles. The study of occupational factors also has a special place in cancer epidemiology.[145] While the Kennaways in their publications of 1936 and 1947 looked for, but did not find a clear association between certain occupations and the striking increase in lung cancer deaths in England and Wales, there were clusters of malignant disease that had long been associated with certain occupations or even certain regions. It was well known, for example, that the miners extracting metal ores in the region around Schneeberg in Saxony and St Joachimsthal (Jachymov) in Bohemia were susceptible to respiratory problems and rarely lived to old age. Physicians working in these areas used a wide range of labels for the respiratory problems that affected the miners and in many cases killed them prematurely: 'suffocative catarrh', 'miners' asthma', 'metallic asthma', 'mountain asthma', 'dry asthma', 'mountain peripneumonia', 'miners' phtisis', 'mountain sickness', 'metal disease', 'toxic pulmonary consumption' and 'miners' exhaustion'.[146] In the late nineteenth century, this specific miners' disease was subjected to the new methods of laboratory medicine discussed in Chapter 2, redefined according to the new concepts of cellular pathology; it was identified as a form of lung cancer.[147] Another occupation affected by lung cancer was asbestos workers. By the late twentieth century, asbestos was the leading cause of occupationally-related cancer death and the second most fatal manufactured carcinogen.[148] But the numbers of lung cancer deaths linked to tobacco smoke are greater by orders of magnitude than the local clusters of occupational lung cancers in the Czech-German border region, and even those attributable to asbestos.

With hindsight it may be tempting to follow those like Horace Joules who thought that the breathtaking rise in lung cancer figures since the 1920s was a scandal. If any infectious disease had been found to kill its victims in those numbers, the consequence would have been alarm and upset. The fact that the public response was comparably tame is instructive. While lung cancer might have been viewed as a self-inflicted disease, it was not caused by, say, dubious sexual exploits, but by a habit generally seen as respectable, which was indulged in by a majority of men and growing numbers of women. We should not judge politicians too harshly who failed to implement convincing policies against smoking, even though the evidence appears so clear with hindsight. As Cole observed, 'the bigger the vested interest the higher the standard of proof demanded'.[149] In a democracy, government responses are never clear cut. The interests involved with smoking were diverse, leading to strange alliances: industrialists and share holders in the tobacco industry joined forces with trade unionists and Labour MPs fearing for workplaces or defending the main pleasures of the working classes. The majority of voters indulged in the habit, knowing that they were putting their health at risk well before the cigarette packages carried official warnings.

A growing number of doctors, however, were willing to blame tobacco not only for lung cancer but also for a range of other illnesses. I will discuss the impact of the association of lung cancer with smoking on the treatment of the disease and clinical research in the next chapter.

5
Trials and Tribulations: Lung Cancer Treatment, circa 1950 to 1970

In the early 1950s, as I have shown in the previous chapter, lung cancer and its possible association with the smoking of cigarettes was an issue of much public debate. I will now turn to the treatment of lung cancer around the same time. How was it organized? As discussed in Chapter 3, surgeons had grown confident by around 1950 that they could help lung cancer patients. In this chapter I will examine how and why their confidence turned into disillusionment within a decade. I will look at therapeutic research and experimental treatments for lung cancer: a series of trials overseen by the MRC, demonstrating that lung cancer at this point was not neglected by researchers, as today is sometimes suggested. I will then discuss the increasing sense of crisis among thoracic surgeons in the 1960s, when progress appeared to be stalling, leading to the feeling among many of them that, as lung cancer was a particularly recalcitrant disease, and in light of the association with smoking, the emphasis needed to be on prevention rather than treatment.

A royal malady

King George VI was probably the most prominent patient in Britain subjected to the new surgical methods discussed in Chapter 3.[1] Like many other heavy smokers the king had suffered from respiratory problems in the past. In 1951 his programme of public engagements had to be curtailed due to problems with his blood circulation, for which he also underwent an operation. In spite of the reduced workload he continued to be very tired and his appearance gave rise to concern. The king himself blamed his bad health on the 'incessant worries & crises through which we have to live'.[2] During a ceremony in

93

Westminster Abbey on 24 May, he looked ill. He had insisted on going through with the ceremony despite a slight fever and in the evening retired to bed with what was thought to be influenza. As he did not make the quick recovery that was hoped for, his doctors used the occasion for a thorough examination of his condition. They X-rayed his chest and the photograph showed a shadow in the left lung. They thought that this was due to a catarrhal inflammation and treated him with daily injections of penicillin for a week, leading the king to believe, as he wrote in a letter to Queen Mary, that 'this condition has only been on the lung for a few days at the most so it should resolve itself with treatment. ... Everyone is very relieved at this revelation & the doctors are happier about me tonight than they have been for a week.'[3]

The doctors were less happy than he assumed. When the king did not recover, a prolonged period of convalescence was prescribed. He spent June and July at the Royal Lodge in Windsor and the Sandringham country retreat in Norfolk, and on 3 August moved to Balmoral Castle in Aberdeenshire, Scotland. His condition improved and, according to his biographer, John Wheeler-Bennett, he 'could enjoy a whole day of shooting without undue fatigue', until again, he 'developed a chill and a sore throat'.[4] His doctors were summoned to Balmoral on 1 September, and upon their request he returned to London on 8 September for further examinations and a tomography of the king's thorax (a series of X-ray images taken from different angles), which suggested the presence of a tumour. His physician, Sir Horace Evans, requested a consultation with Sir Clement Price Thomas, thoracic surgeon at the Brompton Hospital. They agreed that the king should undergo a bronchoscopy. This was a rather unpleasant procedure during which the surgeon inserted a metal tube to the patient's windpipe, which allowed the optical examination of the bronchus and the removal of a tissue sample that could then be examined by a pathologist.

Price Thomas undertook the bronchoscopy in Buckingham Palace on 16 September. 'Only those who were near him', according to the surgeon's obituary in the *BMJ*, 'realized the strains imposed upon [the king]'.[5] The examination of the tissue sample confirmed what the tomography had suggested: George VI was suffering from lung cancer. His doctors held another conference and agreed that there was no alternative to surgery, despite the king's circulation problems and the resulting risk of him suffering a thrombosis. Price Thomas told him about the operation, but without mentioning that he had cancer. He was merely told that a blockage in the bronchus made it necessary to remove one of his lungs. According to Wheeler-Bennett, the king was, 'in a sense, consoled by the thought that the ill-health which had irked him all summer might now be relieved permanently'.[6] 'If it's going to help to get me well again', Wheeler-Bennett quotes His Majesty, 'I don't mind but the very idea of the surgeon's knife again is hell'.[7] On 18 September the following public announcement was issued:

> During the king's recent illness a series of examinations have been carried out, including radiology and bronchoscopy. These investigations now show structural changes to have developed in the lung. His Majesty has been advised to stay in London for further treatment.[8]

Three days later, on 21 September another bulletin announced the imminent operation. The pneumonectomy was performed on the morning of Sunday, 23 September by Sir Clement Price Thomas in spite of the considerable risk of another thrombosis. There was also a possibility that damage to nerves controlling the larynx might affect the king's ability to speak. All went well, however, and another bulletin was issued in the afternoon:

> The King underwent an operation for lung resection this morning. Whilst anxiety must remain for some days, His Majesty's immediate post-operative condition is satisfactory.[9]

The king regained his strength, but only slowly – which should not surprise readers, given the severity of the operation. He seems to have been aware of this. In mid October he wrote to his mother, Queen Mary that:

> At last I am feeling a bit better after all I have been through in the last 3 weeks. I do seem to go through the most serious operations anybody can do, but thank goodness there were no complications & everything has gone according to plan. I have been most beautifully looked after from the surgeon to the nurses & the doctors. They have done their best to make me feel as comfortable as possible. I have been sitting up in a chair for the last week and have had my meals up as well. So I am getting stronger and can walk to the bathroom. It will take some time for me to recover from the ordeal I have been through. ... I must now start to get up and do more to get stronger. Always an ordeal to begin with as one does not know how much one can do on one's own.[10]

During the king's convalescence, Clement Atlee's Labour Party lost the general election and Churchill, aged 76, received the mandate to form a government for the third time. On 30 November George VI left Buckingham Palace for the first time since his operation, to spend the weekend at Windsor Castle. Sunday, 2 December was declared a Day of National Thanksgiving for the king's recovery. Meanwhile he developed a troublesome cough, and Price Thomas performed another bronchoscopy on him. The cough disappeared, but a certain hoarseness remained, which was apparent in his Christmas broadcast. According to Wheeler-Bennett: 'A great sense of peace and happiness now descended on him.'[11] In the new year he took up shooting again at Sandringham (with the approval of his doctors) 'and was delighted to find that he achieved his normally high standard'.[12] On 5 February he shot hares 'with his usual accuracy' and was 'as carefree and happy as those about him had ever known him to be'. He planned the next day's sport, was 'relaxed and contented' at dinner, and retired to his room at 10:30, where he took care of personal affairs until around midnight. After a peaceful night, in the early morning of 6 February, he died unexpectedly. The official cause of death was a blood clot in the heart.

Joint clinics and specialization

His Majesty was not a normal patient and received far more medical attention than the average lung cancer sufferer. A BECC survey of cancer in London revealed in 1952 that out of 1,024 lung cancer patients only 178 were operated on, and only one of these patients survived the operation for five or more years.[13] Out of the same patient population, 239 were treated by radiotherapy, and only four of these survived five years or longer. Six hundred and seven were treated neither by surgery nor radiotherapy. A typical lung cancer patient around 1950 was a middle-aged man who would see his GP about chest problems: pain, a persistent cough, breathing difficulties or even blood in the sputum. The first task of the GP was to decide whether or not a patient was suffering from one of the 'conventional' chest problems, such as tuberculosis or bronchitis. As lung cancer was still rare compared to these illnesses, it was quite likely that a carcinoma remained undiagnosed for a considerable time or indeed was never identified correctly at all. Among a group of patients diagnosed with the disease between 1944 and 1948 and seen at a lung cancer clinic run jointly by the Brompton Hospital and the Royal Cancer Hospital, for

28 percent the symptom that triggered their visit to the doctor was a persistent cough, for 20 percent it was pain, 14 percent had coughed up blood, 9 percent suffered from breathlessness, 6 percent from a febrile illness, 3 percent from general listlessness, and 4 percent from a combination of these symptoms.[14] Patients waited an average of two to three months before consulting a doctor. Thanks to what the radio-therapist David Smithers called 'a revolutionary improvement in diagnosis', mostly due to the wider availability of X-ray facilities and procedures such as bronchoscopies, it was probable that significantly more cases of lung cancer were identified correctly in the 1950s than in earlier decades.[15] Smithers suspected, though, that the accuracy and reliability of diagnosis still left much to be desired. For nearly 10 percent of the 553 patients referred to the joint clinic with a diagnosis of lung cancer by 1948, the diagnosis was rejected after further examinations; an illustration of the difficulty of diagnosing the disease correctly.

In an ideal case, if a GP suspected anything serious, if for example a suspected pneumonia failed to respond to treatment with antibiotics, he or she would refer a patient to the local chest X-ray service. If a shadow was visible in the X-ray, and no tuberculosis bacilli in the sputum, cancer was a possibility. A sputum sample would be screened for malignant cells and the patient would be referred to a chest surgeon for a bronchoscopy.[16] At the Brompton Hospital in the early 1950s, an average of three bronchoscopies were performed per working day, compared to 28 in the whole of 1930.[17] If there were malignant cells or anything suspicious was observed through the bronchoscope, and if the patient's general condition was sufficiently good, the surgeon would schedule an exploratory thoracotomy, or immediately a pneumonectomy or lobectomy (I have discussed these operations in detail in Chapter 3). A focus of innovation since the 1940s, therefore, apart from the operations as such, was diagnosis, the search for methods that allowed the distinction between operable and non-operable cases.

Was a patient indeed suffering from lung cancer, his or her chances of survival were rather small, even if this was diagnosed correctly. Out of the London population sample of 1,024 lung cancer patients studied by the BECC Survey in 1952, only five patients survived their diagnosis for five years or longer.[18] Forty percent of all patients received treatment, either by surgery or radiotherapy. These figures indicate that more patients were treated for their lung cancers in the 1950s than in earlier surveys – and more in the metropolis than in the provinces.

Basil Mackenzie estimated in 1939 in a study commissioned by the Ministry of Health and covering the whole country, that then only 5 percent of lung cancer patients received any treatment at all.[19] Only for liver and pancreatic cancer the rates of untreated patients were higher. Mackenzie was not surprised about these figures as it was 'generally accepted', as he put it, that treatment of cancer at these sites was 'not practicable except under special circumstances; removal of or interference with these organs is, generally speaking, incompatible with life'.[20] Interestingly, Mackenzie's report was published six years after Evarts Graham's acclaimed first pneumonectomy for lung cancer and only four years prior to the production of the British Council film which I discussed in Chapter 3, demonstrating that surgery then was far from standard treatment for this disease in Britain. Another interesting observation was that for stomach and intestinal cancers the rates were almost as low as for cancer of the lung, which Mackenzie thought remarkable as for these cancers surgery was 'a well-recognised more or less accepted line of treatment'.[21] With the exception of gynaecological cancers and those of the breast, it was the rule rather than the exception that cancers were left untreated. Ten years later, in 1949, the Deputy Director of the Holt Radium Institute in Manchester, Margaret Tod found that in the Yorkshire industrial town of Bradford, 87 percent of the lung cancer patients registered at the Regional Radium Centre still had not received any treatment, either surgery or radiotherapy.[22] According to Smithers, who was director of the radiotherapy department at the Royal Cancer Hospital and radiotherapist to the Brompton Hospital, even among the fairly select group of lung cancer patients referred to the joint lung cancer clinic at these hospitals between 1944 and 1949, there were 24 percent for whom he and his colleagues could not do anything, 'despite our anxiety to try at least to relieve symptoms where possible', and in only 28 percent of cases they attempted a radical treatment by surgery or radiation.[23] But was this cause for depression? According to Smithers:

> Your view of the effect of treatment on patients with lung cancer depends (like most things in life) on the way you look at it. If you see over 13,000 deaths in 1951 on a steeply rising curve, and estimate that probably fewer than 250 pneumonectomies for lung cancer were performed in that year, that it is unlikely that more than 1,000 radical courses of x-ray treatment were given for this disease, and that the five year survival rate for the whole group is likely to be no more than a fraction of 1%, you may well feel hope-

less about the situation. If you see an extremely difficult problem both in early diagnosis and in treatment in an often rapidly fatal disease which is at last being tackled with some success, you may take a more optimistic view. The fact is that at present we offer no effective treatment at all to most patients with lung cancer.[24]

Smithers and the physician at the Brompton Hospital, Maurice Davidson, along with the thoracic surgeon J. E. H. Roberts had launched the joint clinic in 1944 as a reference clinic for patients from both hospitals who were suspected of having lung cancer, for treatment by radiotherapy and later also chemotherapy, to find out how useful these therapeutic modalities could be for this particular group of patients. The plan was that suitable patients should be seen weekly at the Brompton Hospital. Any member of the medical or surgical staff of either hospital could refer cases to the joint clinic and take part in its consultations. If surgical treatment was deemed necessary, the patient would be admitted to the Brompton. If X-ray treatment was indicated, the patient would be taken into the Royal Cancer Hospital as soon as a bed was available. A few cases would be treated as outpatients, if their condition allowed them to travel daily from their homes to the hospital.[25]

The patients referred to the joint clinic were indeed a select group, even among the patients treated at the hospitals concerned. More than 400 of the patients diagnosed at the Brompton with lung cancer between 1944 and 1948, for example, were not referred.[26] Some of these were found suitable for surgery and operated without referral. Others were evidently so ill that a referral was thought undesirable. Smithers considered the selectivity of the patient group a major handicap for achieving the clinic's aim, to reliably assess the results of treatment. Another problem for the researchers was that most of the patients died at home and few post-mortem examinations were made. However, by the 1950s, Smithers and his colleagues were granted access to the records of all patients treated for the disease at both hospitals, even when the patients themselves were not referred to the joint clinic.

The joint clinic was both a new, team-based approach to specialization and a clinical research programme; careful record keeping and follow-up were integral to the exercise.[27] Smithers' ideas regarding the reorganization of lung cancer treatment appear to have been met with interest at the Brompton Hospital, especially from chest physicians who were equally research-minded. Davidson and Roberts co-founded the joint clinic with Smithers; it also involved the physicians William Donald Wykeham Brooks, Kenneth Robson, John Guyett Scadding and

John Reginald Bignall, and the surgeons Clement Price Thomas and William Paton Cleland.[28] Scadding, whose role as organizer of the MRC lung cancer trials I will address in the next section, is hailed as one of the founding fathers of modern respiratory medicine in Britain; he was one of the founders and later president of the Thoracic Society and the first editor of the journal *Thorax*.[29] Along with members of the MRC tuberculosis research unit he was involved in running the much-hailed MRC streptomycin tuberculosis treatment trials, which contributed to establishing the randomized controlled trial (RCT) as the new gold standard in clinical research.[30] He was also a member of the investigating committee that prepared the Report on Smoking and Health published by the Royal College of Physicians in 1962 (Scadding himself gave up smoking in 1945).[31] John Reginald Bignall, who became the Secretary to the working group organizing the trials, was consultant physician at the Brompton Hospital and later Dean of the Institute of Diseases of the Chest. From 1956 to 1978 he also edited the journal *Tubercle*.[32] Bignall compiled a book on *Carcinoma of the Lung* published in 1958 as the first volume in a series edited by Smithers, *Monographs on Neoplastic Disease at Various Sites*.[33]

Much of the clinical part of Bignall's book was based on data and experiences from the joint clinic. This included sections on the course of the disease, the effect of various factors on survival and on the different treatment modalities. The conclusions were sobering. The course of lung cancer, Bignall found, was governed by three main factors: the malignancy of the tumour, its situation in the lung and the age of a patient. All three of these were uncontrollable, making this disease appear particularly recalcitrant:

> The chances of our influencing the disease are at present small. Certainly, early diagnosis increases the chances of successful removal or destruction of the lesions; but the difficulties of diagnosing lung cancer early enough are so great that it is highly improbable that any major alterations in the course of the disease can be brought about until some way is found of halting or reversing the unknown changes in the body that make cancer malignant.[34]

The book also included a brief section on the manner of dying, about which, as Bignall contended, 'little has been recorded'.[35] Information about the final stages of the illness had to be collected from GPs and other hospitals as only a few of the patients died at the Brompton. However, it was clear that the more distressing forms of death, such as fatal haemorrhages, ulcerating skin metastases, or effects of large tumours such as choking or dysphagia had become rare, thanks to radiotherapy. About two-thirds of the patients died slowly with increasing weakness and wasting.[36] In only a few cases was there unbearable pain from the primary tumour or from metastases, but it was rare that there was no pain at all. I will return to these issues in the next chapter.

While the patients treated at the joint clinic were a select group and by no means representative of lung cancer patients in Britain, the rigorous record keeping and follow-up procedures championed by Smithers and his colleagues, and similarly in other centres such as the Christie in Manchester, provided opportunities to collect data on treatment success for this disease. However, the longer the follow-up, the clearer it became that the results were not particularly good news for lung cancer sufferers. In 1967, based on 6,086 cases, Bignall, Smithers and their co-authors reported a five-year survival rate of 20–35 percent in patients having a resection. There were few long-time survivors following irradiation, a path taken when patients were deemed unsuitable for surgery. The available therapeutic measures proved 'sadly ineffective', they found, both because of the recalcitrant nature of the disease and the late stage at which it was usually treated.[37] About two-thirds of the patients seen with lung cancer died within the first year after attending hospital, and the five-year survival rate in Britain was probably little more than 5 percent. The article reflected the disillusionment that characterized the late 1960s and to which I will return towards the end of this chapter.

We will now leave the Brompton-Royal Cancer Hospital joint clinic, but stay with some of the main protagonists. There was a well established network of clinicians with expertise in lung cancer in existence in Britain by the 1950s, including Smithers and his colleagues associated with the joint clinic, who contributed to the MRC cancer trials to which I turn in the next section.[38]

The MRC treatment trials

On 31 January 1957, five months before the publication of the 'Statement on Tobacco Smoking and Cancer of the Lung', the Medical Research Council held a Conference on the Evaluation of Different Methods of Cancer Therapy.[39] The conference, under the chairmanship of the Professor of Radiotherapy at Middlesex Hospital Medical School, Brian Windeyer, recommended that the Council 'should consider undertaking an investigation into the treatment of certain tumours which appeared particularly suitable for short-term study'.[40] They included in this group carcinoma of the bronchus, oesophagus and bladder, bone sarcoma and medulloblastoma. Carcinoma of the bronchus was chosen 'after considerable discussion' because of 'the vast amount of material which was available and the existence of a good deal of confusion of thought about the best form of treatment'.[41]

The notion that lung cancer seemed comparably well understood followed from the intense interest generated by the rise in the incidence of this once rare and obscure disease. As I noted in the previous chapter, clinicians and pathologists had observed more and more cases since the beginning of the twentieth century, and debates over whether this increase was real and over its possible causes grew more intense towards the 1950s. Surgery by this time was the standard treatment, but the recommendations of the 1957 conference were heavily geared towards the evaluation of new approaches in radiotherapy – the form of therapy from which British cancer specialists most expected innovative impulses, in spite of disappointments with the treatment of lung cancer.[42] This was perhaps not surprising, given the strong presence of radiotherapists in the committee. The MRC had played a central part in the rise of radiotherapy in the UK in the interwar years, and radiotherapists were increasingly discontent with the key position of surgeons in the treatment of malignant disease.[43] They rebelled against the notion that surgery was the default treatment for all patients who had any hope of survival, while they found it difficult to recruit trial subjects. Radiotherapists increasingly perceived of themselves as generic cancer specialists, while surgeons specialized on organs or parts of the body.

In June 1957, the Council appointed a Steering Committee to prepare the appropriate trials. The members of the Steering Committee were Brian Windeyer (chair, London), the radiotherapist Joseph Mitchell (Cambridge), the physician Robert Hunter (St Andrews), the pathologist Robert Scarff (London), the surgeon Alphonsus d'Abreu (Birmingham), the pathologist Jethro Gough (Cardiff), the epidemiologist and statistician Austin Bradford Hill (London) and the haematologist Leslie Witts (Oxford).[44] The Steering Committee appointed five ad-hoc working parties to plan trials for each of the chosen forms of cancer: carcinoma of the bronchus, oesophagus and bladder, bone sarcoma and medulloblastoma.[45] But the plans met with little enthusiasm. Three of the working parties were disbanded in short order: the carcinoma of the oesophagus working party (chair: D. F. Campbell) immediately after its first meeting, as its members agreed that the envisaged trial of radiotherapy versus surgery did not make much sense. The working parties for carcinoma of the bladder (chair: R. B. Hunter) and medulloblastoma (chair: Dorothy Russell) both terminated in 1962. The bladder cancer group did not manage to recruit enough patients fulfilling the admission criteria for the planned trial and subsequent proposals were deemed not practicable. The medulloblastoma group gave up when they found, after undertaking retrospective studies, that comparable cases treated with usual procedures in different centres could not be matched to obtain a statistically significant evaluation of the results of treatment. Only the working parties for carcinoma of the bronchus (chairman: J. G. Scadding) and bone sarcoma (chairman: Herbert Seddon) survived until the early 1970s, by which time the MRC fundamentally reorganized its activities in cancer research, partly due to generational change (see Chapter 6). A sixth, chemotherapy-focused working party appointed in 1955 under the Oxford haematologist Leslie Witts to work on Leukaemia (a move partly made in response to the success of similar studies underway in the United States) fared the best of all. The leukaemia working party carried out a number of trials in leukaemia and myelomatosis, and in 1968 was reconstituted as an independent, permanent Council Committee. Without looking at the particular circumstances it is hard to tell why some working parties succeeded and others did not. Their composition followed the same pattern: each included a physician, a surgeon, a pathologist, a radiotherapist and a statistician. In the case of lung cancer, however, the working party included research-minded physicians (and to a lesser degree surgeons), who were collaborating anyway (for example in the London joint clinic).

The research programme drawn up by Windeyer's committee was at least as much about the development of new methods of clinical research as it was about finding new therapies for cancer. With the initial proposals of all working parties in hand, the committee recommended that: 'Favourable consideration should be given, if possible, to the support of any suitable clinical trials in the field of cancer therapy which can be carried out without too elaborate an organization and with reasonable promise of yielding useful information.[46] The committee was a vehicle towards applying the new RCT approach to the evaluation of well-established and new therapeutic methods, especially in radiotherapy. The use of this new approach for establishing the effectiveness of streptomycin in the treatment of tuberculosis had provided the MRC with a much-publicized success.[47] Extending the Council's activities to cancer research (the domain of the Imperial Cancer Research Fund and the British Empire Cancer Campaign) was part of an MRC strategy to establish the Council as the main body controlling clinical research in Britain.[48] The RCT approach, which built on some of the key strengths of MRC-funded clinical research and came to embody its ethos, was a vehicle for this strategy.[49]

The members of the Working Party appointed to assist the Steering Committee with the organization of the trials on carcinoma of the bronchus were, besides Scadding, the radiologist Leslie G. Blair (London), the surgeon Alphonsus L. d'Abreu and the pathologist Jethro Gough, as well as Bradford Hill and Windeyer. Along with Bradford Hill, who was consulted whenever the Council needed statistical expertise, Scadding had also been involved in the streptomycin trials.[50] The Working Party recommended the Committee undertake a randomized trial of different forms of radiotherapy in a small number of centres (they explicitly mentioned Edinburgh, Newcastle, Manchester, Liverpool, the Middlesex Hospital and Hammersmith Hospital). Reconstituted for this purpose, under the same Chairman, the Working Party was to prepare and oversee the trial. In 1959, Scadding's colleague at the Brompton, John Bignall was appointed as secretary to the trial.[51]

Ethics and feasibility

It is interesting to look at the preparation and organization of the MRC lung cancer trials in some detail as it tells us much about the ways in which therapy was locked into a paradigm – surgery wherever possible and as early as possible – which made it difficult to consider alternatives. This situation was made particularly precarious by an increasing feeling of frustration about lung cancer treatment among surgeons, which I will discuss later in this chapter. It is also important to note that randomized controlled trials did not always come natural to cancer researchers, as we might assume if we only look at the history of blood and lymph or childhood cancers, and especially if these trials involved treatment modalities other than chemotherapy.[52] The discussions among both the Steering Committee and the Working Party organizing the lung cancer trials centred predominantly on what kinds of studies were both technically and ethically doable. It turned out that the two realms, the technical and the ethical were difficult to keep separate. Ethical concerns, for example, were frequently raised over the prospect of randomization, an issue that was central to the Committee's work. One of its members, Professor Scarff wondered 'if strict randomisation was necessary since so many clinicians had a clear-cut impression of what was best for the patient and might feel random selection to be a little unethical'.[53] Bradford Hill, seconded by Mitchell and Hunter argued that randomization was in fact necessary in order to detect marginal differences. To Hunter this 'raised in its train the question of feasibility again'.[54]

How were such problems to be overcome and appropriate trials organized? And why did the Committee choose to look at lung cancer therapy? Hunter was concerned that 'there were many forms of cancer which could not be suitably used in such an investigation because the pattern of treatment was so well established and so widely accepted that any deviation would cause ethical difficulty, and that this left free for investigation only the fringe of cases of hopeless prognosis'.[55] Windeyer disagreed, arguing that there were cancers, such as carcinoma of the bladder, for which several forms of therapy were successfully used, but where confusion existed over the relative merits of the different treatment regimens. The committee members viewed lung cancer as particularly suitable as its high incidence and short natural history (after diagnosis) promised large numbers of trial subjects in a reasonable time span. Otherwise they appeared to view carcinoma of the bronchus as a representative problem rather than a recalcitrant exception.

The committee agreed that retrospective surveys could not supply the answers they were looking for. Long-term studies were too expensive (and this was where a short natural history was useful), but at least five years follow-up were necessary. However, not only survival should be recorded. Other parameters were also to be taken into account, such as time spent out of hospital, time spent out of work, degree of pain and disability, dyspnoea and haemoptysis. For lung cancer it was especially important, Mitchell suggested, 'to evaluate the ordeal of treatment against possible benefit, and to try to decide if, in the late cases, X-ray treatment was worthwhile as opposed to simple palliation'.[56]

But this is where the problems started. At a meeting in 1959 the Working Party found it almost impossible to define criteria that distinguished palliation from prolongation of life, and defining criteria was an important stage in the organization of trials.[57] It became increasingly obvious, as I will argue below, that ethical difficulties were integral to the whole enterprise of organizing clinical trials for cancer therapies.

Finding a suitable question

Soon after the constitution of the ad-hoc working parties in 1957, it became clear that it was not easy to find a suitable, well-contained question, which could be answered by way of an ethically acceptable clinical trial, within the remits set by the Recommendations (promising, reasonably easy to organize, using randomization, and leading to further research). While the motivation for the trials partly derived from the streptomycin success, American cancer research also served as a model. P. Armitage of Hammersmith Hospital was invited to report on experiences with co-operative, multi-centre trials in the United States and he also discussed a trial in progress at Hammersmith, comparing surgery with radical radiotherapy in operable cases, but with a very limited intake of patients – an attempt to compare different methods of radiotherapy had failed for technical reasons.[58] Scadding suggested three problems that might fit the remit of the recommendations and were worth studying, first, the efficacy of surgery as opposed to radiotherapy, 'which was as yet an unsolved question', second, the efficacy of different kinds of radiotherapy, and third, the use of chemotherapy alone or in combinations with other forms of treatment. However, he did not believe that there was satisfactory evidence for the beneficial effects of chemotherapy, and therefore he did not think that an evaluation of its use was a suitable subject for an MRC trial. There were also, he argued, considerable ethical objections to a comparison of surgery and radiotherapy, as nearly a quarter of the patients undergoing surgery survived for five years or longer. Scadding, for these and other reasons, was sceptical about the Hammersmith trial.[59] The Working Party concluded that, while desirable, 'a large-scale controlled investigation of the relative merits of surgery as opposed to radiotherapy did not appear feasible at the present time'.[60] The main factor that made such a study appear unfeasible was the expectation that it would be difficult to obtain the necessary co-operation of surgeons.

The discussions in the committee seemed to go in circles and progress was frustratingly slow. Since so far only 'some 10 percent' of patients were considered for therapy at all, Windeyer asked, would it not be possible to study the remaining 90 percent, maybe by comparing different forms of radiotherapy?[61] Representing the surgeons, who were more interested in improving diagnosis, d'Abreu argued that what was 'badly needed' was information about the relative prognosis in different kinds of cancer of the bronchus, and whether the prolongation of life by a few months by means of radiotherapy was worth the price patients paid in terms of quality of life. Finally they came to a conclusion about which nobody was really enthusiastic: 'that a comparative trial of different methods of radiotherapy might be considered in patients not primarily suitable for surgical treatment but regarded suitable for an attempt at cure by radiotherapy'.[62] The inclusion of chemotherapy was 'not thought to be practicable at the present stage of knowledge in this field'.[63] The details of the trial, however (and this was an indicator of the increasing frustration), were to be determined by a working party with a different constitution.

The Working Party decided to approach Philip D'Arcy Hart of the MRC's Tuberculosis Research Unit about co-ordinating the work in collaboration with the Statistical Unit at the London School of Hygiene and Tropical Medicine, because of these units' previous experience with controlled trials.[64] It appears as if the Working Party hoped that they were going to be able to repeat their streptomycin success. D'Arcy Hart, initially reluctant because of staff shortages in his unit, accepted the offer and appointed a new member of staff, Dr Joan Heffernan.[65] Letters were written to the centres that the Working Party considered as likely participants in the study and a joint meeting was prepared with radiotherapists.

It was important for the Working Party to persuade a sufficient number of radiotherapists to collaborate, so it was important to involve them in the preparation. A meeting with 29 consultant radiotherapists took place on 21 January 1961 in the Council Room of the Royal College of Surgeons in London.[66] The Chairman told those present that '[c]onsiderable difficulty had been encountered in making plans which would not only be ethically acceptable and feasible but which, at the same time, could produce information of value'.[67] The purpose of this meeting, therefore, was 'to find out whether the radiotherapists concerned were in agreement about the importance of the principle of controlled clinical trials and whether further agreement could be reached upon a subject worth trying and upon the methods involved'.[68] Would the radiotherapists provide the Working Party with clearer directions? They did not. The radiotherapists, too, were unenthusiastic about the proposals. Some had doubts if a trial in carcinoma of the bronchus made much sense in the first place. Ralston Paterson from Manchester conceded that 'some difficulties were implicit in random selection, but he hoped that radiotherapists would encourage the Medical Research Council to continue to organize a trial'. However, he suggested that 'lung cancer was one of the more difficult fields for investigation as overall mortality is high and it is difficult to assess differences in response'.[69] Others proposed that a trial should be designed to 'compare the progress and survival rate of patients with presumed undifferentiated carcinomas of the lung following surgical treatment, with that following radiotherapy'.[70] The Working Party followed their suggestions, and I will now look at the results.

As depressing as it was predictable: The MRC trial comparing surgery and radiotherapy

Four years after the decision to organize lung cancer treatment trials the frustratingly slow negotiations over the details of these trials appeared to draw to a conclusion. Another consultation meeting was scheduled with both consultant surgeons and radiotherapists on 25 July 1961.[71] Scadding introduced the agenda by stating that 'there appeared to be a clinical problem as to the right advice to give a patient with a histological report of an undifferentiated carcinoma of the bronchus – whether to advise surgery or radical radiotherapy'.[72] Defining the problem in this way helped to overcome ethical problems: 'For those who honestly felt they did not know which treatment to advise there were no ethical difficulties. If there were enough people with this doubt in their minds, the trial could be conducted.'[73] Gwen Hilton at University College Hospital had reported results with radiotherapy in a small number of cases, which were 'apparently as good as surgery'.[74] The discussion with the surgeons, moreover, indicated that there was indeed disagreement. While some of the surgeons saw it as proven that resection, where possible, was always superior to other forms of treatment, others argued that for this kind of tumour it was time to move away from surgical treatment as the results were uniformly poor, and turn to radiotherapy or chemotherapy. In most places, according to one radiotherapist (Dr Fleming, St Thomas), only 'surgical rejects' were treated with radiotherapy.[75]

A central problem was eligibility of patients for the study. Following the consultation with the radiotherapists, the Working Party had returned to a trial plan that its members had dismissed at an earlier stage of the discussion, but now restricted to what some described as 'anaplastic', others as 'undifferentiated' carcinoma of the lung.[76] Restricting eligibility on grounds of cell type made a study feasible that originally was unacceptable on ethical grounds to some members of the Working Party. But tumour grading was a difficult business. One of the surgeons present at the meeting with the Working Party (Mr Nohl,

Harefield) pointed out that in his experience nearly one-fifth of histological reports were mistaken.[77] Could it be so difficult, the reader might ask, to recognize a cancer cell under the microscope? In fact, recognizing cancerous cells was and still is not easy. The distinctions are gradual. The less a cell looked like normal cells found in a particular organ and the more individual cells varied in size and shape, the more likely this was cancer. And the more cells on the slide under the microscope were in the process of dividing, the more likely the tissue sample was cancerous. Furthermore, grading schemes had changed significantly since the first attempts to classify tumour cells in the nineteenth century, with new techniques and pathological material becoming available, and later for pragmatic reasons, when new forms of therapy encouraged further distinctions between cell types. Some tumours proved to be more susceptible to certain treatments, which made distinctions meaningful that had not carried any meaning before.[78] In the early 1960s, a subgroup of 'undifferentiated' or 'anaplastic' bronchial tumours were reclassified as carcinomas of small cell or oat cell type (the terms were used interchangeably).[79] This reclassification exercise was partly driven by experiences with chemotherapy (these tumours had proved responsive to cytotoxic chemicals in early trials) and partly by attempts to establish an internationally consistent terminology.[80] By the time the first results were published in 1966, the trial was described as 'Comparative Trial of Surgery and Radiotherapy for the Primary Treatment of Small-Celled or Oat-Celled Carcinoma of the Bronchus'.[81]

The results of the trial were not encouraging. After two years, only three of the original 71 surgical patients and ten of the 73 radiotherapy patients were still alive. According to the report in the *Lancet*, '[t]he number of survivors at 24 months is so small that further statistically significant differences between the series in this respect cannot now arise.'[82] Both policies had produced very poor results. The working party suggested that radiotherapy may be the slightly better choice, as post-operative complications would be avoided.

> However, because the results of the treatment are so poor whether by surgery or radical radiotherapy there is an urgent need for further research to improve the treatment of this condition. There is also an urgent need to apply the knowledge already available, in particular that of the role of cigarette smoking ... to the *prevention* of the disease.[83]

We can see how in the light of increasingly wider acceptance of the tobacco hypothesis the focus was shifting towards prevention. A note in the administrative file dealing with the study states:

> It seems to me that there is nothing at all controversial about this report, which is a straightforward account of a difficult but well organized clinical trial, the outcome of which has been as depressing as it was predictable.[84]

Nevertheless, follow-up for the 13 survivors continued as planned. After five years, only one surgery patient and three in the radiotherapy group were alive, and after ten years the last surgery patient had also died.[85] The three survivors in the radiotherapy group were still alive and well.[86]

The official goal of this and other clinical trials was to provide unbiased evidence that would lead to closure in a controversy. The debate unfolding on the letter pages of the *Lancet* after the publication of the first report showed that this was not achieved. The study was criticized by leading pulmonary surgeons. Roger Abbey Smith of the Thoracic Unit at the King Edward VII Memorial Chest Hospital in Warwickshire argued that the reason why the results for surgery were so bad was that only patients with particularly unsuitable centrally located oat cell tumours were included in the study.[87] Nine of the 71 patients in the series had not undergone surgery because their condition had deteriorated too rapidly, and of 58 patients where exploratory surgery was performed, 24 had been found to be inoperable. In all patients whose tumours had been surgically removed, this had been done by pneumonectomies rather than less risky lobectomies. Abbey Smith argued that the results for surgery would have looked much better had peripheral tumours been included, and the Working Group could therefore only claim validity for centrally located oat cell carcinomas. John Rashleigh (Jack) Belcher, a friend of Abbey Smith's based at Middlesex Hospital doubted if the conclusion that radiotherapy was superior to surgery in this situation was valid.[88] Kent Harrison (St Thomas's) worried that uncritical readers might wrongly conclude that surgery had no place in the treatment of small cell carcinoma. He believed that any apparently operable carcinoma of the lung should be removed surgically, no matter what the cell type, and for this reason had not participated in the trial.[89] One US surgeon still criticized the trial in the 1980s for the inadequate image of hopelessness he thought it had created around lung cancer surgery, especially for oat cell carcinoma.[90]

Scadding defended the study that his Working Party had organized, arguing that even taking these criticisms into account, the results were not significantly different and the outlook remained bleak:

> The facts should be publicised: the incidence of a disease which has assumed epic proportions, which has a high mortality, and for which no current method of treatment can be regarded as satisfactory, would be reduced to a small fraction of its present level if men and women as responsible individuals chose to give up, or never to take up, cigarette smoking.[91]

It seems that the frustrating outcomes of a trial about which nobody was very enthusiastic in the first place, reinforced an ongoing shift of emphasis from therapy to the prevention of lung cancer. However, by presenting experiences with the treatment of small cell carcinoma (an especially vicious type of tumour), as representative of lung cancer more generally, it may be argued that Scadding made the outlook for lung cancer patients seem bleaker than necessary.

The second trial overseen by the Working Party was a study of chemotherapy as an adjuvant to surgery. Like the first trial, this second one was also co-ordinated by the MRC Tuberculosis Research Unit.[92] The preparation of the chemotherapy trial, it seems, was much smoother than that of the first study. There were no extensive debates among the Working Party and no big meetings with consultants. One explanation for this lack of controversy may be that chemotherapy was only tested as a secondary therapy, an adjuvant to surgery, to prevent the growth of secondary tumours. It may also be due to the fact that unlike radiotherapy, in chemotherapy there were few entrenched positions. It was perceived as something new, an approach that promised new channels for intervention (and also something that the British were not particularly good at and still needed to learn).[93] However, while the Working Group had shown that it was able to organize a clinical study in lung cancer that conformed to the new standards of a co-operative, double-blind, randomized controlled trial, the results had little practical value: 'The therapeutic results at two years are disappointing, for there is no evidence that either of the two cytotoxic drugs in the dosage used improved survival ...'[94] After five years, 27 percent of the patients who received cyclophosphamide were still alive, 28 percent of those on busulphan, and 34 percent in the placebo group.[95]

Mass miniature radiography and lung cancer

Could the persistent difficulties to treat lung cancer successfully be due mostly to late detection and, indeed, was a solution already available, in form of a screening campaign using the existing infrastructure for the detection of tuberculosis? [96] A 1952 textbook stated that 'mass X-ray examination of the chest offers the best known single means of detecting, as rapidly and economically as possible, significant intrathoracic lesions in presumably healthy subjects'.[97] In contrast with tuberculosis screening, there was no epidemiological benefit to be gained from the detection of lung cancer, as this was not a communicable disease. The results were 'purely personal'. But if early diagnosis improved the prognosis for these patients, would this not justify a national campaign for periodical chest X-ray examination of 'persons in the critical age groups'?[98] During 1955, mass radiography units in England and Wales detected 1,482 cases of lung cancer.[99] During the same year, a total of 17,272 deaths from lung cancer were recorded, suggesting that fewer than 10 percent of new cases were picked up by mass radiography. Furthermore, of the positive cases, almost half were not detected in routine mass radiography screenings but in special sessions for symptomatic cases referred by general practitioners. Clearly, there seemed to be room for improvement.

Only very little information was available in the early 1950s, however, on the fate of patients whose lung cancers had been picked up in mass radiography screenings. E. Posner, Medical Director of the Stoke on Trent Mass Radiography Unit, L. A. McDowell, Advisor in Mass Radiography to the Birmingham Regional Hospital Board, and K. W. Cross, a medical statistician, began in the summer of 1955 to systematically collect data on patients in the Birmingham area, whose lung cancers had been detected by mass radiography, either in routine surveys or after referral by their GPs, to get a clearer idea about the usefulness of this technology for the detection of this disease.[100] The results of their study were ambiguous. On the positive side, Posner, McDowell and Cross found that a (slightly) higher proportion of the tumours detected in routine surveys proved to be resectable, and more of the operations were lobectomies rather than more debilitating pneumonectomies, indicating that cancers were less advanced. The survival rates were also higher for this group. Out of 93 cases detected in routine surveys before 1957, 44 patients underwent lung resections, and 20 survived for five years or longer (38 percent of the operated patients and 6 percent of those where surgery was not possible). This compared to 138 out of 402 cases referred by GPs who were operated, and 45 who survived for five or more years (27 percent of the operated patients and 3 percent of those not operated). The better survival figures for cases picked up in routine surveys, however, could not be taken at face value. It was likely that these included a higher proportion of slow-growing cancers (the higher survival figures for untreated patients pointed in the same direction). The slower the cancer grew and, thus, the less harmful it was, the higher the chance that it was noticed in a routine survey.[101]

The proportion of 'truly silent' lesions detected in routine surveys, moreover, was rather small. Interviews with the first cohort of patients revealed that only 15 percent of the men whose cancers were found in this way had really not noticed any symptoms. This low detection rate suggested that the technique was simply not sensitive enough. Even experienced observers failed to see the shadows of small tumours located on the periphery of the lungs on the small 35mm films. After 1957 the 35mm cameras were replaced with 70 and 100mm units.[102] However, this led only to a small increase in the detection rate in routine surveys, from 0.3 per thousand to 0.4 per thousand.[103] This means that they now found one lung cancer case among 2,500 men they screened, compared to one case in 2,900 men with the older technology. The low detection rate raised the issue of costs. How expensive was it to screen the population for lung cancer? The organizers of the Birmingham study in 1963 calculated that it cost almost £1,000 to find one case of lung cancer in routine mass radiography. This compared to an estimated £20 per tuberculosis patient.[104] They concluded, therefore, that 'orthodox' routine mass radiography, that is, periodical visits to factories and other localities at three- to four-yearly intervals had 'little to offer in the way of finding lung cancer with favourable prognosis'.[105]

Conclusion: Almost unmitigated gloom

All these disappointing results added to growing frustration among chest surgeons and contributed to a general sense of gloom among those who dealt with lung cancer. In the introductory address to a symposium on the diagnosis and treatment of carcinoma of the bronchus in 1966, the chest surgeon Norman R. Barrett suggested that: 'The contribution that surgeons can make to this subject, in so far as the solution of the problem is concerned, has already been made, and I do not believe there is much more any surgeon can add with advantage.'[106] After evaluating the results of several decades of surgery for bronchial carcinoma, thoracic surgeons now were at a stage where 'the *raison d'être* should be discussed – why we are doing surgery at all'.[107] He viewed the idea of a 'cancer operation', the radical removal of a tumour and its lymphatic field in one block, as suspect. Just because it worked for breast and rectal cancer, this did not mean that the concept was applicable to lung cancer. Neither did he think that diagnosis would make a difference – 'Is there any evidence that improved methods of diagnosis would help the results of surgery? I think there is none' – or that early detection was the answer: 'I would suggest to you that the conception on which surgeons have worked – namely that if they are sent the patients early their results will be good – is nonsense.'[108] Further tinkering with surgical procedures was pointless: 'I think the technique of surgery in relation to immediate survival after operation for carcinoma of the bronchus is standardized, effective and efficient.'[109] Barrett did not stop with surgery: 'Radiotherapy in the treatment of malignant disease has played its part' and '[c]onventional pathology also has little more to offer'.[110] New solutions, Barrett argued, could only be expected from laboratory scientists. Surgeons had to get up to speed with modern cancer research and biology rather than hope that improvements of technique were going to make any difference. Surgeons, however, Barrett suggested, while not likely to conduct any further promising research themselves, still played an important role as clinicians who helped laboratory researchers with careful observations of tumour development and its clinical manifestations, 'in recording the exceptions rather than the rules'.[111]

Barrett was not alone with his thoughts about the end of progress in thoracic surgery. P. R. Allison of Oxford confessed to similar doubts in an address dealing with 'The Future of Thoracic Surgery', delivered to the Society of Thoracic Surgeons of Great Britain and Ireland in October 1965. 'Carcinoma of the lung is treated by surgery now as well as it ever will be treated by surgery', he argued. 'This is a technical exercise, but the ultimate solution to carcinoma of the lung must surely be a biological one and not a surgical one.'[112] Neither was this attitude unique to Britain. In the discussion following Barrett's opening speech at the Midhurst symposium, J. S. Chapman of the South Western Medical School, University of Texas, agreed with the speaker and added that '[t]he point of view Mr Barrett has expressed this morning is almost identical with that of many other thoughtful surgeons in the United States'.[113] Jack Belcher characterized the predominant mood during the symposium as 'almost unmitigated gloom'.[114]

The gloom also extended to questions of diagnosis. The radiologist G. Simon admitted that even on full scale radiographs it was difficult to detect tumours early. The smallest tumour he ever managed to identify on a routine radiograph, he reminisced, was about 1.3 cm big. 'In terms of cellular proliferation', he commented, 'a lesion this size can hardly be called "early"'.[115] If they were obscured by shadows of the ribs, by the heart or by other organs in the chest, tumours as large as 2 cm could easily be missed. And only about 25 percent of the lung was completely unobscured in a normal chest X-ray of an adult man.[116] Such problems led the chest physician at the Brompton Hospital and Secretary of the lung cancer working party, John Bignall to comment on the issue of early detection as follows:

> With the existing procedures early diagnosis is still largely a matter of chance. More patients would be diagnosed early if they sought advice earlier, but in many cases symptoms are so insidious and unobtrusive that the disease will inevitably be at an advanced stage before this occurs. The value of frequent routine chest x-rays has not yet been demonstrated. ... The immediate outlook for improving early diagnosis is not good. Nor is the immediate outlook for improving the results of treatment. It is therefore all the more necessary to try harder to prevent this largely preventable disease, which is so difficult to detect in the early stages of its evolution.[117]

As I discussed in Chapter 3, surgeons had developed the first promising operations for this once rare disease in the 1930s. Since then they had 'owned' lung cancer, controlled the procedures and overseen the progress of patients.[118] Opening the thorax had confronted surgeons with a formidable challenge and, as Pack and Ariel put it in their celebratory account of *A Half Century of Effort to Control Cancer*, '[t]he chest was one of the last anatomic regions to be surgically invaded'.[119] Overcoming these problems, with great difficulty and with the help of modern technology, had been a source of pride and professional power. When depression about the stagnation in lung cancer surgery overcame them in the 1960s, thoracic surgeons increasingly saw heart surgery as the new challenge, as the field where fame and power could be gained. Around the same time, chest physicians assumed a somewhat more central role in the diagnostic pathway, which they cemented with the introduction of the flexible fibreoptic bronchoscope in the early 1970s, operated by physicians more frequently than surgeons.[120] As I will discuss in the next chapter, medical oncologists also developed an interest in lung cancer and adapted chemotherapy regimens for this disease, but there was no major breakthrough, and the mainstay of the curative (as opposed to palliative) treatment of non-small cell lung cancer remained surgery.[121]

The difficulties that the organizers of the MRC trials encountered show that using any modality other than surgery as a first-line therapy was ethically questionable, which made it almost impossible to organize clinical trials comparing different therapies, unless they were adjuvant to surgery or targeted patients who some frustrated trial organizers termed 'surgical rejects'. In a rare trial in the mid 1950s, involving the other established locally acting treatment modality, radiotherapy (rather than systemic, like chemotherapy), Gwen Hilton, radiotherapist at University College Hospital London, treated patients suffering from operable lung cancer and, in fact, obtained results with ordinary deep X-rays (not the newest and most powerful machines at the time) that were very similar to those that the surgeons achieved.[122] Ten years later, during the gloom of the Midhurst symposium, her co-author Joseph Smart, physician at the London Chest, the Brompton and the Connaught Hospitals reminisced:

> The problem ... was to get the cases. When I suggested the investigation to Dr Hilton she said she would like to do it but she never got these cases. This was very understandable as any surgeon seeing an operable carcinoma naturally has to take it out.[123]

However, that the results of radiotherapy equalled those of surgery indicated that the form of therapy had only a limited effect on the course of the disease and long-term survival, supporting an observation that Bignall made in his 1958 monograph. All came down to biology: the malignancy of the tumour.

Not all thoracic surgeons agreed with the pessimistic outlook of Barrett and Allison. Dissenting voices could be heard even in the deep gloom of the Midhurst meeting. Jack Belcher confessed that after one and a half days of listening to gloom he 'felt increasingly proud to be a surgeon since we at least can do something for this disease'.[124] Belcher, it seems, and a small number of colleagues, maintained their enthusiasm until they retired in the 1980s, by which time they were veterans of their trade.[125] Around the time of his retirement, Belcher published long-term follow-up results of three decades of pulmonary surgery performed by himself and other members of Charlie's Club, a small, informal society of thoracic surgeons that he convened.[126] These data showed that survival rates after surgery changed surprisingly little from the 1950s, when lung resections became routine, to the 1980s. Survival rates varied between 25.5 and 26.8 percent after five years, and between 13.6 and 17.8 percent after ten years.[127] Belcher himself recorded a five-year survival rate of 28 percent for the period from 1950 to 1955 and of 27 percent for 1970 to 1975, while the ten-year survival rate remained unchanged at 15 percent. The rates were similar for conservative surgeons and for their more aggressive colleagues and it did not make any difference for survival if lobectomies (resection of parts of a lung) or pneumonectomies (resection of a whole lung) were performed.[128] To Belcher the 'disease process itself' seemed to be the dominant factor that determined survival rates.[129]

There was, in fact, some progress that could be recorded since the 1950s. The results that Belcher and his colleagues reported were significantly better, for example, than what the BECC survey of cancer in London revealed in 1952, when out of 1,024 lung cancer patients 178 were operated on and only one patient survived the operation for five years or more.[130] Most London surgeons then had obviously not reached the high standards of specialists such as Belcher and his colleagues, and this observation seems to vindicate the move towards specialization as it was advocated, for example by Smithers. Out of the same patient population, 239 were treated by radiotherapy, and only four of these survived five years or longer. Six hundred and seven were treated neither by surgery nor radiotherapy. Especially the surgery figures make the survival rates achieved by Belcher and his colleagues but also by the Brompton surgeons look very desirable. What made the difference, it seems, were the expertise and routine achieved by specialist chest surgeons, along with effective referral pathways. The challenge at this stage was administrative and political rather than technical. How could as many lung cancers as possible be diagnosed early (and correctly), referral processes streamlined, those who would benefit from surgery identified reliably, and operated on in a specialist chest unit fairly quickly. And while there was no reliable cure and a smoking ban was unrealistic, how could as many smokers as possible be persuaded to quit?

In the next chapter I will deal with the rise in the 1970s and 1980s of another treatment modality, chemotherapy, which went along with a generational change in cancer treatment and cancer research and the adoption of the protocol approach in lung cancer trials. I will also look at the development of international standards and the adoption of the TNM classification system, which made both the assessment of patients and the exchange of data easier.

6
More Enthusiasm, Please: Preventing, Screening, Treating, Classifying, circa 1960 to 1990

By the late 1960s lung cancer had turned into the most common form of cancer worldwide, with a death rate almost as high as the death rate from all other forms of cancer combined.[1] Not only could there be little doubt that smoking was not good for smokers' health (it said so on cigarette packs from 1967 in the US and from 1971 in the UK). It was also becoming increasingly clear that once diagnosed with lung cancer, patients had very limited hopes of surviving the disease. And it did not look as if this was about to change. In this respect developments for lung cancer were strikingly different especially to childhood cancers, where the outlook for patients had been dire in the 1950s and which were increasingly viewed as curable with new chemotherapy regimes.[2] Patients diagnosed with lung cancer not well enough for surgery, with tumours in awkward locations or cell types known as so malignant that an operation was going to be futile, ran out of options very quickly. To be sure, lobectomies or pneumonectomies on lung cancer patients still constituted much of the daily work of thoracic surgeons, but this was not the most rewarding work. The low survival rates could be depressing, and as we heard in the previous chapter, surgeons expected little change. Results for radiotherapy or chemotherapy were equally disappointing. But what could be done about this disease?

Could lung cancer be prevented? Smoking remained a major concern among public health experts. As Virginia Berridge has shown, public health was changing since the 1950s, driven partly by changes in society more broadly.[3] The 1970s saw, in Britain, the launch of a vocal anti-smoking lobby, acting with government support: the pressure group ASH (Action on Smoking and Health). I will briefly review its early history in this chapter. But what about those who had already developed lung cancer? Since the 1950s, some surgeons had argued that lung tumours were simply not diagnosed early enough. If they were picked up earlier, they suggested, more patients would be able to benefit from surgery. I will discuss debates about early detection in this chapter. Was it possible and worth the effort to screen for lung cancer, especially among those at increased risk (smokers above middle age), considering that there was mass screening infrastructure for tuberculosis in place?

I will also return to the MRC and clinical cancer research. Not only did the 1970s witness a generational change, with the cohort of Smithers, Scadding or Bignall retiring, who had shaped lung cancer research in the 1950s and 1960s. There was also a change of direction. The new trials mostly involved forms of chemotherapy, adopting new style of practice that Keating and Cambrosio have described.[4] Finally I will look at what chest surgeons viewed as the most significant change in the treatment of lung cancer over the last few decades: given that there was no hope for a miracle cure it was important to identify with greater precision those patients who would benefit from the limited range of treatment options. This was made easier by an internationally standardized classification and staging system, which not only facilitated more streamlined treatment pathways, but also aided the adoption, for lung cancer treatment, of that new style of practice associated with large-scale, sometimes international, co-operative clinical trials.

Routines and realities

Before turning to the changes of the 1970s and 1980s, let us take stock and reflect on what a lung cancer patient could expect at the beginning of this period. One major difference compared to the early 1950s was a heightened awareness of the disease and its major cause, cigarette smoking. By the late 1960s, people in Britain, the US and elsewhere had been exposed to more than a decade of health education telling them that smoking was dangerous and caused lung cancer and other illnesses. Furthermore, other respiratory illnesses, so abundant in the early 1950s, were becoming rarer. As a consequence, it was more likely that a case of lung cancer was indeed diagnosed correctly.

But was it not already too late when there were symptoms? 'Silent' cases of the disease were occasionally picked up during routine exams such as mass radiography campaigns. If a shadow was visible on a chest X-ray, there was still the possibility that the underlying lesion was some kind of infection. But if the shadow did not disappear after a short and intensive treatment with antibiotics, it had to be assumed that it was malignant. In some such cases the presence of cancer could be confirmed by way of a bronchoscopy or by analysing cells in the sputum, but often only an operation could provide certainty.[5]

Surgery, the removal of the tumour, was still the form of treatment that gave the best long-term results. Radiotherapy delivered good palliation and relief from some of the unpleasant symptoms of advanced lung cancer, but did not produce 'cure' rates that were comparable to those achieved with surgery. The use of cytotoxic drugs – chemotherapy – was becoming more common, but it was a continuing matter of controversy if there was any use to it. Lung cancer specialists, however, knew much more about the influence of different cell types on the response to treatment. Chest surgeons increasingly agreed that 'oat' cell or small cell carcinomas were rarely worth operating as they spread very quickly and were likely to have already done so by the time of diagnosis. The reluctance to operate appeared to be vindicated by the results of the MRC trials I discussed in the previous chapter.[6] Tumours of a second cell type, adenocarcinomas, were often easy to operate, but patients rarely survived their operations for very long. Patients diagnosed with cancers that belonged to the third and most common cell type, squamous-cell tumours, had the best chances of survival. Many were suitable for surgery and if none of the glands were affected, more than half of these patients could expect to survive the removal of a lung for five years or longer.[7] The position of a tumour in the chest and the age of the patient were also important. But as cancer specialists always emphasized, every tumour was different and a law to itself, and decisions could be very subjective. While younger patients were often fitter, their tumours were also often more malignant. Patients over 60 or 70, on the other hand, occasionally survived well beyond the average 18 months that patients could expect to live following their diagnosis if they did not receive any treatment. For many of these older patients surgery was not an option.

Out of the more than 25,000 Britons diagnosed per year with lung cancer in the mid to late 1960s, only about 20 to 25 percent were found to be fit for an operation, and in only 15 to 18 percent the removal of the tumour proved possible in the event. Of these, one-third survived for five or more years. Thanks to greater awareness, improved assessment and better surgical techniques, the proportion of patients who died as a consequence of the operation, 15 to 20 percent in the early 1950s, had been reduced to 5 to 10 percent. By 1980 about 12 percent of the men and women presenting with primary lung cancer were referred to surgery. Out of these, 80 percent were found, during the operation, to have resectable tumours (The higher selectivity of referral was an effect of the developments I sketch in this chapter).[8] If patients were operated successfully, they were usually out of bed within two or three days and walking after a week. They had to expect that their breathing was affected, and their ability to be physically active. As the eminent chest surgeon Sir Thomas Holmes Sellors put it: 'A reasonably fit middle-aged man, able to play tennis before operation, will have to satisfy himself with golf; a healthy young adult may notice little change, but an elderly barrel-chested chronic bronchitic may suffer serious limitation and have to restrict his mode of life considerably.'[9]

In cases where surgery was not an option, there was the possibility of radiotherapy. However, the results were often disappointing. While treatment with X-rays alleviated symptoms and in some cases led to the regression of a tumour, complete cures were rare. And there were also the side effects such as the loss of appetite, sickness and malaise that were suffered by many patients, and these were difficult to endure, especially for patients whose lung cancers had not yet caused any clinical symptoms. Chemotherapy, according to Holmes Sellors, 'had no real place in treatment' apart from 'considerable psychological value'.[10]

What could be done for patients for whom no 'active' treatment with the objective of a cure was possible? The most distressing features of terminal lung cancer were shortness of breath and cough, and pain if the cancer had spread to the bone or the chest wall. When moderate doses of standard cough depressants failed to bring relief or give patients enough rest to sleep, Holmes Sellors saw no reason to withhold morphine or morphine derivatives. He recommended the classic, so-called 'Brompton Cocktail' or 'Brompton Mixture', which contained morphine and cocaine and 'has proved valuable in the terminal stages in relieving symptoms and inducing euphoria which avoids a great deal of distress to patients and relatives'.[11] 'The more fortunate patients', according to Holmes Sellors, were 'those whose growth is centrally placed in the lung field or in whom liver or adrenal secondaries produce a fairly symptomless downhill course'.[12]

Death and dying

The majority of patients in the 1960s appear to have ended this downhill course at home, and many without ever being told that it was lung cancer that was killing them. In 1963 Eric Wilkes, then a general practitioner in Baslow, Derbyshire (later a leading member of the British palliative care movement), undertook a survey of cancer patients who died in the Sheffield conurbation.[13] Out of 3,422 patients who died from malignant diseases, 1890 (55 percent) died at home. Wilkes sent a questionnaire to all family doctors who certified such deaths and asked them for further information about up to two cases each. He collected details on 374 cases, 20 percent of those who died at home. Lung cancer patients formed the largest group within this sample. Lung cancer was also the commonest form of malignant disease seen at the pioneering St Christopher's Hospice. More than 35 percent of the male and 8 percent of the female cancer patients cared for in the hospice in the early 1970s were suffering from terminal lung cancer.[14] The age distribution in Wilkes's Sheffield sample was, as Wilkes put it, 'as expected', with nearly half of the patients over 70 years old. Of the 80 lung cancer patients 67 were men and 13 women. For 75 percent of these patients the doctors who responded to Wilkes's survey found the conditions at home satisfactory (the figures were similar for the other main cancer sites).

This left, of course, about 25 percent where the conditions were bad. The comments by GPs included the following examples: 'A difficult and inadequate wife', 'An attitude of doom and helplessness', a 'Terrible home. Refused hospital', or 'Next of kin unconcerned. Lodger cared for him'.[15] In 27 percent of all cancer cases the GPs classified the medical care that the patients needed as 'minimal', for 59 percent as moderate, and for 12 percent as heavy. Ninety-three percent were cared for by relatives, and 63 percent of the patients were (also) visited by the district nurse (again, the distribution was similar for all main cancer sites). Fifty-two percent of the lung cancer patients died without a period of difficult nursing, in 26 percent there were less than six weeks of difficult nursing, and in 14 percent the period was longer than six weeks. Only 13 percent were found to need hospital care. 'The pattern of illness revealed by this survey is surprisingly constant', Wilkes concluded. 'Half the cancers of lung or stomach, breast or bowel, or other sites have a smooth journey with little in the way of pain or suffering.'[16] Only three out of the 80 lung cancer patients (compared to, for example, nine out of 37 patients with breast cancer) were told that it was lung cancer that sent them onto this journey. This may indicate that the diagnosis was considered too depressing to share with a patient. Or it may point to a stigma associated with this disease, more so than other cancers. I will return to this issue in the next chapter.

While the majority of deaths today occur in hospitals, palliative care has become firmly established as part of the medical mainstream since the 1970s, especially for cancer. The hospice movement launched by Cicely Saunders, while not necessarily associated with one particular religious faith, had strong spiritual roots, and I will return to the spiritual aspects of cancer charity in Chapter 7.[17] In Britain, organizations such as Macmillan Cancer Care and Marie Curie Cancer Care train and supply specialist nurses who support relatives caring for dying cancer patients at home. Marie Curie Memorial was founded in 1948 by committee members of the Marie Curie Hospital in Hampstead, following the hospital's transferral to the NHS.[18] Their objectives were to provide a nursing and welfare service for patients in their own homes, along with the provision of nursing homes and hospices, and funds for 'the encouragement of scientific learning'. In 1952 they funded a survey of cancer care at home, jointly with the Queen's Institute of District Nursing.[19] The findings of the survey formed the background for much of the charity's work over subsequent decades. Macmillan started as a small, local organization in 1911.[20] The budgets and activity portfolios of both charities expanded greatly in the 1970s. The income of Marie Curie Memorial increased from less than 4 million pounds in the period from 1967 to 1971, to nearly 20 million pounds between 1977 and 1981. Macmillan funded their first Macmillan nurses and built their first Macmillan Cancer Care Unit in 1975. By the 1980s they were funding nursing teams throughout the UK, as well as training programmes for doctors and nurses.[21]

Smoking and the new public health

Virginia Berridge has argued that much changed in public health in the 1970s, in terms of attitudes to tobacco and in the relationships with government.[22] The 1950s and 1960s were characterized by an attitude that Berridge characterizes as 'Systematic Gradualism' – the belief that much could be achieved if government, public health experts and the tobacco industry collaborated, for example in the search for a less harmful, 'new smoking material'. Tobacco manufacturers were seen as partners. The 1970s, according to Berridge, was a period of change in the ideology of public health, featuring an increasingly central role for expert committees as an instrument of policy making, an increasingly technocratic breed of public health experts, and, crucially, the rise of anti-smoking activism, as exemplified by the pressure group ASH, from 1973 under its new director, Mike Daube. The changes went along with a new 'absolutism' in attitudes towards smoking. The goal was, increasingly, to make smoking socially unacceptable.

In the 1970s, there was not significantly more media coverage of smoking and health than in earlier decades. A quick survey of newspaper databases shows that in both tabloids and broadsheet newspapers the issue probably received more attention in the 1950s, in the wake of the findings by Doll and Hill and subsequent government statements, than in the 1970s. But the status of smoking was changing in the 1970s. Cigarette smoking was no longer as deeply embedded in social and cultural practices as in previous decades. There was by now a broad consensus among experts that the habit was a cause of lung cancer and other diseases. While in 1967, still about 75 percent of British men smoked at least occasionally, smoking was now increasingly depicted as not only stupid and a waste of money, as suggested by a Central Office of Information poster campaign in the 1960s, but also as somewhat deviant, damaging to others, and irresponsible.[23] By the time passive smoking became a major issue after 1981, when the results of a Japanese study suggested that the non-smoking wives of heavy smokers faced an increased risk of developing lung cancer, the notion was starting to take hold that smokers not only damaged their own health but also the health of others.[24] This strengthened the hand of anti-smoking campaigners and changed the direction of public health policies.

ASH and its campaigns exemplified these changes. The organization was launched in 1971 and funded mostly by the UK government. The inspiration, according to Sir George Godber (chief medical officer in the Department of Health and Social Security from 1960 to 1973), came partly from the US, where relationships between tobacco industry and government were increasingly adversarial.[25] ASH's function was to be that of 'a thorn in the flesh', an anti-smoking lobby – a role that epidemiologists and public health experts shunned as many felt uncomfortable with the necessary partisanship it would bring with it. The organization was meant to continue and take further the developments that the MRC and the Royal College of Physicians had started with their statements on smoking in 1957 and 1962, keeping smoking and health issues in the public arena and continuing to put pressure on the government, parts of which were reluctant to support effective anti-smoking policies. The initiative came from Charles Fletcher, the secretary and driving force of the committee that had authored the RCP Report, and Keith Ball a doctor based at the Central Middlesex Hospital, home to many of the champions of social medicine in Britain and the MRC Social Medicine Research Unit.[26] Initially the new organization was medically dominated. Its campaigns were relatively low-key and often directed towards other doctors, until in 1973 a new director was appointed. Mike Daube came to ASH from the charity Shelter, a pioneering, new-style campaign organization against homelessness and squalid housing conditions founded in 1966. Daube was not attracted to the job at ASH because he had particularly strong feelings about smoking, but because he felt that this was 'a pressure campaign that was ripe'.[27] 'You had your villain', he reflected in an interview, revealing what he thought made for a good campaign: 'You had your St George and the Dragon scenario, you had your growing ecology bandwagon, growing interest in consumerism. It seems there were a lot of prospects of making something out of it.'[28]

Daube chose campaign tactics for ASH that were designed to achieve effects rather than educate; he was a spin doctor. As he told Berridge in another interview, Daube followed an American activist text, *Rules for Radicals*: 'rule one is to personalise the problem – the people running the major companies are responsible for those deaths'.[29] The organization purchased individual shares in tobacco companies, allowing them to expose and attack company representatives (now clearly viewed as villains) during Annual General Meetings. They also collaborated with journalists like Peter Taylor, who in the 1970s produced anti-smoking television programmes with titles such as *Dying for a Fag* (the profile of a man in his early forties dying from lung cancer and talking about his fatal love affair with cigarettes), *Licensed to Kill* (a film about the tobacco industry) or *Death in the West – the Marlboro Story* (a group portrait of six real-life Marlboro men, American cowboys at various stages of dying from emphysema or lung cancer).[30] ASH lobbied MPs and encouraged the planting of parliamentary questions. Why would the government provide the main funding source for a thorn in its flesh? According to Berridge, ASH fulfilled a facilitating role in the political process that was viewed as useful. She argues that 'its media profile was part of its attraction to politicians because it made it appear as a "force to be reckoned with" and therefore useful as a counterweight to the stance of industry and of other government departments'. ASH shows for the United Kingdom how the tools of consumerism could be used against the tobacco companies, contributing to the transformation of smoking into a habit that was viewed as not only unhealthy but also somewhat reckless and anti-social. Anti-tobacco campaigns succeeded in turning the tobacco industry into a pariah, in the US even more than in Britain. They also contributed to the stigmatization of smoking, and by implication, of illnesses that were known to be caused by the habit. While other forms of cancer – notably breast cancer – have become de-stigmatized since the 1970s, lung cancer probably carried more of a stigma by the 1990s (I will revisit these issues in Chapter 7).[31] Two decades after the first publications by epidemiologists on the link between cigarettes and lung cancer, changing attitudes towards smoking were gradually taking effect. By the 1970s lung cancer death rates among men reached a plateau and by the 1980s they were falling.

Lung cancer and the gospel of early detection

It is widely assumed that the next best thing to prevention is early detection (also known as 'secondary prevention').[32] This appears to be common sense: if cancer is diagnosed early, before it grows too far or spreads to other organs, surgeons can operate and radiotherapists radiate tumours more effectively, and patients have better chances of surviving the disease.[33] As Robert Aronowitz has shown, this has generally been held to be the case for breast cancer since Halsted developed the radical mastectomy, even though Halsted himself never promoted the 'Do Not Delay' message.[34] The early detection message has been promoted relentlessly by breast cancer campaigns in the US.

If early detection is desirable, the conceptual leap towards screening risk populations, developing programmes that identify a disease before it causes symptoms, is relatively small.[35] At first glance, lung cancer was the ideal disease for a screening programme. The high-risk group could be defined relatively easily – cigarette smokers above a certain age, say, 50 – and in many places health authorities could draw on an existing service: mass radiography screening for tuberculosis (see Chapter 5). A lung cancer screening programme had the potential of becoming the first programme directed at a disease that affected predominantly men.

Historians of cancer screening have so far focused predominantly on women's diseases such as cervical and breast cancer.[36] Screening campaigns for these female cancers have become models for others. The Papanicolaou (Pap) smear for detecting precancerous cervical cells came into wide use in the United States in the late 1940s and early 1950s. By the 1960s its sensitivity in detecting cervical cancer in its preclinical stage was considered to be nearly 100 percent.[37] If positive smear results were confirmed by standard diagnostic methods such as biopsies, women were usually referred to a surgeon to have the suspicious lesions removed. The wide use of the Pap smear, however, has led to a somewhat paradoxical development. On the one hand it appears to have contributed to a significant decrease in the mortality from cervical cancer. On the other hand it is associated with an increase in the incidence of the disease. This problem appears to be common to many screening programmes. Such programmes, as their promoters stress, have identified many early cases of cervical cancer in time for successful treatment. However, as critics argue, many clusters of anomalous cells identified as precancerous lesions and surgically removed, would never cause trouble if left alone. Screening programmes, thus, while saving lives, also turned women into cancer patients who would never have developed clinical symptoms.[38]

While cervical cancer screening programmes were associated with a decline in mortality, this was less clear for breast cancer. With its reliance on radiological diagnosis, mammography screening is more directly comparable to lung cancer screening. Mass screening for breast cancer started in the 1960s when mammography became widely available. It remained controversial, however, despite a series of large-scale trials, whether these programmes actually reduced breast cancer mortality, and if they did, whether the overdiagnosis associated with screening and the distress caused by a diagnosis of cancer for a lump that never would have turned into a problem, did not outweigh the benefits.

The so-called Health Insurance Plan (HIP) study was a much-discussed randomized control trial of breast cancer screening, roughly comparable and probably serving as a model to later lung cancer screening studies in the US. In December 1963, its organizers enrolled 62,000 women members of the HIP of Greater New York aged between 40 and 64 years for this trial, who were randomly assigned to a study and a control group. Women in the study group were invited to have a screening examination consisting of a mammography and a physical examination, usually by a surgeon. If no suspicious lesion was found, they were offered three annual follow-up examinations. Those in the control group were not screened but were entitled, obviously, to all other HIP benefits, including, if and when they explicitly requested them, general physical examinations. The study organizers were looking for a difference in mortality between the study and control groups. The results after 18 years of follow-up appeared to provide the supporters of mammography screening with evidence for their case.[39] However, while the breast cancer death rate was 23 percent lower in the intervention group than in the control group, critics argued that the HIP study really tested the combined effect of mammography and breast exam, demonstrating that women receiving the clinical exam and mammography did better than those receiving standard 1960s medical care.[40] In Britain, a trial in Edinburgh in 1979 did not assign individual patients randomly to mammography and control groups. Rather, the organizers randomized the GP practices that cared for the women. After seven years, women cared for in practices that provided mammography and breast exam were 17 percent less likely to die from breast cancer than those cared for in the control practices.[41] Critics of this study argued that women in the GP surgeries assigned to the control group were on average poorer than in the mammography group, inviting alternative, socioeconomic explanations for the mortality difference.[42] In 1985, Kenneth Clarke, then Minister of Health, convened an expert committee chaired by Sir Patrick Forrest to report on breast cancer screening. The committee came to the conclusion that mammography screening had the potential to lead to a prolongation of life for women aged 50 or older. In 1987 the NHS Breast Screening Programme was established, and since 1988 all women in the UK between the ages of 50 and 64 have been invited to have routine mammographies.[43] The controversy, however, was not over, with critics arguing that for every case of breast cancer that was picked up by routine screening, ten women needlessly underwent surgery, radiotherapy or chemotherapy for lumps that were visible in the mammogram but would never cause them any problems.[44] Moreover, the results of two Canadian studies that started to recruit women in 1980 and followed them for up to 16 years, suggested that women receiving mammography plus breast exams were no less likely to die from breast cancer than those receiving only breast exams.[45] A review of mammography trials by the Cochrane Collaboration in 2001 also found that many of the older studies, whose results had suggested that mammography screening was beneficial, had major flaws.[46]

The example of breast cancer shows how difficult it was to demonstrate conclusively if screening was beneficial.[47] What was the situation like for lung cancer? I have dealt with cervical and breast cancer screening here in some detail because these programmes are often viewed as models. Screening programmes for lung cancer had been considered since the 1940s, as I have discussed in Chapter 5, taking advantage of the mass screening infrastructure established for the detection of tuberculosis.[48] While at first glance it may have looked like a good idea to simply use this infrastructure to also screen for lung cancer, the results of the Birmingham study appeared to demonstrate that this was not viable, especially given that tuberculosis incidence was declining.[49] This still left the option of developing selective schemes for a certain group of the population: middle-aged smokers. But at what age should one start? Results from a study undertaken with six-monthly mass radiography examinations in South London from 1959 to 1963 suggested that for younger men (up to age 54) mass radiography probably did not bring any advantages in terms of life expectancy.[50] In this age group, a high proportion of the tumours were oat celled, and for this cell type the survival rates were especially bad. For male smokers aged 55 or older, a group where both lung cancer incidence was significantly higher and a higher proportion of tumours were of the less malignant squamous cell type, the estimated costs per detected lung cancer went down to £180.[51] But older patients were less likely to be fit enough for surgery. A study carried out by G. Z. Brett at the Mass Radiography Service of the North West Metropolitan Region seemed to indicate that patients under 50 benefitted more from early detection by routine, six-monthly examinations.[52] However, Brett and his colleagues found that while tumours picked up in routine surveys were more likely to be operable, the improvement in five-year survival rates was only modest. The biology of the cancer cells appeared to be more important for the outcome than the time at which a tumour was detected.

Between 1967 and 1969 a Working Party on the Prevention of Cancer of the Standing Sub-Committee on Cancer of the Central Health Services Council's Standing Medical Advisory Committee reviewed the available evidence for and against the usefulness of screening programmes for cancer of the breast, cervix, lung and large intestines.[53] The members of the committee were sceptical about all screening programmes with the exception of cervical cancer. The evidence reviewed on lung cancer included papers by Bignall and Brett, who both were pessimistic about the effect of screening on mortality. Bignall, in fact, was sceptical about the prospect of any intervention for lung cancer.[54] The committee report accordingly gave no reason for hope: while screening by mass radiography could be shown to increase resectability, there was no sign that it improved survival rates significantly. Treatment results were so bad that 'an intensive national campaign for six-monthly radiography could have no more than a trivial effect on national mortality and so cannot be recommended'.[55] The committee concluded that '[t]he only hope for effecting a substantial reduction of mortality from carcinoma of the lung lies in changing the smoking habits of the population'.[56] The benefit of screening for heavy smokers over 55 was 'not sufficient to justify maintaining mass radiography services for this purpose alone'.[57]

Mass radiography services were being scaled down around this time, as tuberculosis was increasingly viewed as conquered, a negligible public health risk. By 1965 the number of deaths per year from respiratory tuberculosis had declined to 2,008 from 20,156 in 1947 and an average of 9,760 in the years 1951–1953.[58] While over the same period the number of deaths from lung cancer had gone up from 9,204 to 26,398, the Labour government under Harold Wilson decided in 1969 to phase out mobile mass radiography services.[59] In summer 1970, when the issue was debated in the House of Commons, there were still 49 mobile and 31 static units in operation in England alone, but regional hospital boards had submitted proposals that would see the numbers reduced to 32 mobile and 29 static units in the following year.[60] By 1978 there were 37 mobile and 19 static units in operation in England and Wales.[61] While the mobile units picked up a fair number of cases of lung cancer (3,783 in England and Wales in 1967) and other conditions (for example 8,841 cases of heart disease), the government obviously did not think they were worth maintaining for screening for non-tuberculous conditions alone.[62] In response to criticism, the government pointed out that stationary radiography services in hospitals were unaffected.

But what about diagnostic techniques with better resolution than mass radiography? In the US, the National Cancer Institute sponsored three randomized controlled trials in the 1970s to evaluate the effect on mortality of screening programmes based on chest radiographs and sputum cytology, as part of the Cooperative Early Lung Cancer Detection Project: the Mayo, Memorial Sloan Kettering and Johns Hopkins Lung Projects. They led to similarly disappointing results as the British studies I have discussed. The Mayo Lung Project was viewed as the most 'definite' of these trials, and the results of the other two were similar.[63] For the Mayo trial, between 1971 and 1976, 10,933 Mayo Clinic outpatients undergoing general medical exams, men over the age of 45 who smoked at least one pack of cigarettes per day, underwent an initial screening consisting of a full-size chest radiograph on 36×43cm film and the cytological examination of a sputum sample.[64] They were then randomly assigned to a study or a control group. The screened group were invited to undergo four-monthly chest radiographs and sputum exams, the control group received no such invitations. However, a majority of them had chest X-rays taken as part of their subsequent medical examinations. In effect the trial compared four-monthly screening with normal Mayo Clinic care (this was similar to the HIP study of breast cancer screening). The results were encouraging at first glance. The screening regime picked up 206 cases of lung cancer while in the control group only 160 cases were confirmed. Lung cancers detected by screening were more likely to be early, operable squamous carcinomas, and the members of the screening group survived longer after diagnosis. Thirty-three percent of the patients whose lung cancers were detected by screening survived for five or more years, compared to 15 percent in the control group. Paradoxically, however, the overall lung cancer mortality in the study group (3.2 per 1,000 person-years) was slightly higher than in the control group (3.0 – the difference was not statistically significant). Later studies undertaken in Czechoslovakia and East Germany echoed the Mayo results; others were inconclusive or showed modest survival benefits in the screened group.[65]

How could the paradoxical results of the Mayo study be explained? To start with, critics found that there were problems with study design and execution – in some ways this is reminiscent of the debate over the quality of the screening trials for breast cancer.[66] More important, however, was perhaps the possibility of overdiagnosis. The argument was similar to the debates over breast cancer and cervical cancer screening programmes: it was likely that screening detected lung tumours that would not have killed patients before they died from other causes. Like other cancers, lung tumours were known by the early 1970s to grow at different rates. While aggressive growths could double in size in about 40 days, others took more than ten times as long.[67] The big difference between lung cancer and both breast and cervical cancer was, however, the link with smoking, a cause of lung cancer that also caused other serious health problems. Many heavy smokers were likely to die from other smoking-related illnesses before their (slow-growing) lung cancers caused them any trouble.

So what were the lessons learned from all these screening trials? The widespread and generally accepted answer to the question if lung cancer screening worked, based on trials and studies from the 1950s to the 1980s, was a fairly resounding no, even if the organizers of the big American trials warned not to completely dismiss screening as pointless. In Britain, interest in screening for lung cancer faded. Mass radiography services were scaled down and screening programmes particularly for lung cancer were never institutionalized. In the 1990s, however, the wide availability of CT scanners gave rise to new hopes and new discussions.[68] Would a different, more sensitive technology make lung cancer screening worthwhile? I will return to this issue in the next chapter.

Chemotherapy and generational change: Clinical lung cancer research in the 1970s and 1980s

With hopes for successful (and cost effective) screening programmes diminishing and surgeons frustrated by the lack of progress in the treatment of the disease, clinical researchers nevertheless did not give up on lung cancer. However, as childhood cancers, leukaemia and lymphomas were increasingly viewed as treatable by way of chemotherapy, the structure and the aims of clinical research were changing. A new generation of researchers were replacing those who had shaped cancer research since the end of the Second World War. Surgery and radiotherapy were no longer at the centre of the enterprise (and definitely not at the cutting edge), while chemotherapy was no longer seen as hopeless. As medical oncologists became established as the youngest specialty at the cancer hospitals in the early 1970s, some of them also concentrated on lung cancer.[69] And the trials that the new generation of clinical cancer researchers organized (not exclusively, but also on lung cancer), increasingly followed the 'protocol' approach described by Keating and Cambrosio.[70]

We can see this development reflected in the history of the MRC clinical cancer research programme overseen by the Committee on Evaluation of Different Methods of Cancer Therapy under the radiotherapist Brian Windeyer since 1957, whose activities up to the late 1960s I have discussed in the previous chapter. By 1969, officials at the MRC headquarters came to the conclusion that this committee had been 'in the doldrums for some time past', while, in contrast, the Leukaemia Committee chaired by Leslie Witts appeared to be thriving and was preparing to set up working parties on Wilms Tumour, Neuroblastoma and Hodgkin's Disease, all malignancies that responded well to chemotherapy.[71] Many of the members of Windeyer's committee had retired or were about to retire, and Windeyer himself felt that the committee was 'moribund' and should be disbanded.[72] As far as trials in radiotherapy were concerned it could be superseded, MRC officials thought, by a new Radiotherapy Working Party set up in 1971 under Norman M. Bleehan, Brian Windeyer's successor as Professor of Radiotherapy at the University of London. Its role was to be 'to initiate collaborative clinical trials in cancer, with particular reference to radiotherapeutic problems'.[73]

The leukaemia trials were already independent of Windeyer's committee, and any policy forming role that the committee may have had, had been taken over by the Co-ordinating Committee on Cancer. Its chairman was the UCH clinician Lord Max Rosenheim, succeeded in 1972 (following Rosenheim's death), by Richard Doll. As discussed in Chapter 5, many of the working parties on specific cancers had been disbanded fairly soon after their inauguration. The lung cancer working party was still going, but according to Julie Neale, the MRC official overseeing the reorganization, 'badly in need of reconstitution'.[74] The working party's chairman, J. G. Scadding was due to retire in 1972 and the group needed 'the injection of more enthusiastic members'.[75] Neale felt personally that this working party 'should be doing far more than it is at present, although this is of course a very difficult clinical area'.[76] At this point in time, clearly, the MRC as one of the main funding bodies of medical research in Britain had no intention of abandoning lung cancer. If there was a shortage of new initiatives, this appeared to be due to a lack of enthusiasm among chest specialists, and their frustration over the recalcitrance of this disease, which failed to respond to all attempts of controlling it.

Who was going to co-ordinate the clinical cancer research programme when Windeyer's Committee was going to be disbanded? If the Radiotherapy Working Party was going to be put in charge of co-ordinating all trials, what would happen if chemotherapy were to become 'a really live issue in other fields'?[77] Bleehen's Committee at this stage did not count any chemotherapy experts among its members. The Leukaemia Committee could continue to stand on its own. If the Cancer Co-ordinating Committee would take on the 'overlord' function, this would encourage greater collaboration with ICRF and BECC, but the Co-ordinating Committee was 'deficient in clinical members'.[78] Finally, in January 1973, the Council decided to disband Windeyer's Committee on the Evaluation of Different Methods of Cancer Therapy and to revise the title, terms of reference and membership of the Radiotherapy Working Party to enable it to take over the functions of the Committee. Bleehen initially proposed to change the working party's title to 'Radiation Therapy-Oncology Working Party', with reference to the US Radiation Therapy Oncology Group, but the Council decided on 'Cancer Therapy Committee'. The proposed new members, chosen in order to broaden the committee's expertise beyond radiotherapy, were the haematologist, pioneer of medical oncology in Britain and Imperial Cancer Research Fund professor at St Bartholomew's Hospital, Gordon Hamilton Fairley, along with

Drs T. A. Connors of the Chester Beattie Institute and Ian Todd of the Christie Cancer Hospital in Manchester, who were both known for work on the chemotherapy of cancer.[79] The Committee was put in charge of advising the Council on all matters related to clinical cancer trials, with particular reference to solid tumours.[80]

Soon after the reconstitution of Bleehen's radiotherapy working group as Cancer Therapy Committee, a renewed and reconstituted Lung Cancer Working Group met in February 1973 at the MRC's London Headquarters in Park Crescent. Scadding had retired in the previous year, and with the exception of the director of the MRC Tuberculosis and Chest Diseases Research Unit, Wallace Fox, none of the members of the 1960s working party were associated with the new group. In terms of members' institutional backgrounds, the group was heavily skewed towards radiotherapy. The new chairman was Norman Bleehen, and he was joined by Thomas J. Deeley, the Director of the South Wales and Monmouthshire Radiotherapy Service and Lecturer at the Welsh National School of Medicine, Cardiff and George Wiernik, Radiotherapist at Churchill Hospital, Headington, Oxford. The chest specialists among the members were Wallace Fox and Ruth Tall of the Tuberculosis and Chest Disease Research Unit at the Brompton Hospital. For the following meeting in July 1973 they were also joined by A. H. Laing, a colleague of Wiernik's at Churchill Hospital, and Ian Sutherland of the MRC Statistical Research and Services Unit. It is worth noting that there was not a single thoracic surgeon among the group.

While the members of the new Working Group were all radiotherapists except the clinical trial and chest specialists Fox and Tall and the statistician Sutherland, the discussions during the initial meetings revolved almost exclusively around chemotherapy. We may want to read this as an expression of frustration with the conventional approach to lung cancer: surgery where possible, to cure; and radiotherapy where necessary, to palliate. To be fair, in both surgery and radiotherapy, radical approaches were giving way to more conservative therapeutic regimes, putting emphasis on the avoidance of hazards and on a good subsequent quality of life.[81] The new openness towards chemotherapy among radiotherapists may also have been part of an attempt to explore possibilities for medical intervention in cases where interventions within the normal repertoire were considered pointless. And it should probably be understood as a response to therapeutic successes with chemotherapy (and the organization of clinical trials involving chemotherapy) in the treatment of leukaemias, lymphomas and childhood cancers.[82] Furthermore, the development reflected the central role of radiotherapists in the United Kingdom as organizers of cancer treatment.[83] Who managed the treatment of lung cancer in practice, depended on the ways in which the disease was framed and on the availability of services. Where lung cancer was viewed primarily as a chest disease it was managed by thoracic surgeons. In locations where major cancer centres existed, however, lung cancer was viewed as primarily a malignant disease and radiotherapists managed treatment pathways which combined different modalities.[84] Bleehen's proposed title for the radiotherapy committee demonstrates that he and his colleagues were keen to rebrand themselves as oncologists, specialists for cancer treatment in general rather than specialists for one therapeutic modality.[85]

Bleehen was as central to the new era of the MRC cancer clinical research programme as Windeyer had been in the 1950s and 1960s. Born in Manchester in 1930, Bleehen grew up in London and studied medicine at Oxford, taking an extra year to do a BA in physiology and biochemistry with the help of an MRC studentship.[86] For his research on aspects of the action of insulin he won a prize. In 1952 he began his clinical training at the Middlesex Hospital Medical School. He qualified in 1955. A house job in the radiotherapy department brought him into contact with Windeyer, and he decided to specialize in radiotherapy. In 1966 he was awarded a Lilly fellowship by the MRC, which allowed him to spend time at Stanford University to work with Henry Kaplan, an innovative US radiotherapist.[87] He was offered a faculty post at Stanford but decided to return to Middlesex. In 1969 he was appointed as Windeyer's successor as Professor of Radiotherapy and head of the Academic Department of Radiotherapy.

In 1975 the MRC invited Bleehen to set up a clinical and research unit at Addenbrooke's Hospital in Cambridge, the MRC Clinical Oncology and Radiotherapeutics Unit. The Cancer Research Campaign endowed a Department of Clinical Oncology at Cambridge University and Bleehen was appointed CRC Professor of Clinical Oncology and Radiotherapeutics, succeeding Joseph Mitchell. Bleehen's main research interests were the treatment of brain tumours, especially gliomas, and lung cancer, both, according to one obituary writer, 'challenging diseases where new treatments were needed to improve dismal prognoses'.[88] He was a founding member of the International Association for the Study of Lung Cancer, member of its Board of Directors and Vice President from 1978 to 1994.[89]

Bleehen was instrumental in the establishment of a new MRC infrastructure for clinical trials of cancer therapies. He believed that the MRC needed a dedicated office to supervise the statistical and data management tasks associated with this approach. A Cancer Trials Office (CTO) was set up within his unit at Cambridge in 1977, with a statistician and data manager, playing an increasingly central role for the MRC cancer trials programme. In 1991 the CTO became the institutional home of the lung cancer team of the MRC Tuberculosis and Chest Diseases Unit. In 1998, the MRC integrated its cancer and HIV trial activities in a new Clinical Trials Unit (CTU) in London.[90]

From its inception, the new Lung Cancer Working Group under Bleehen's chairmanship met more frequently than the old working party had met in the 1960s. If we can believe memos and the minutes of their meetings, the fundamental doubts that members of Scadding's old working party expressed frequently about the value of therapeutic research on lung cancer were absent from the discussions of the new group. Also, as far as one can tell based on memos and minutes, there was more enthusiasm for trying out new methods and treatment regimes, especially those involving some form of chemotherapy. Bleehen himself presented a paper on chemotherapy for lung cancer during a workshop that Keith Ball organized for ASH in 1973, which, admittedly, was 'more concerned with hopes for the future than present achievements'.[91] However, he expressed optimism, justified by success in the treatment of other malignant diseases.

The working group meeting on 23 February 1973 started with a brief stock taking exercise before discussing plans for a randomized control trial comparing non-treatment (or rather treatment delay or placebo) with different radiotherapy and chemotherapy regimens for patients with tumours that were localized but not suitable for surgery (excluding oat cell tumours). Unlike in the 1960s, the 'questions' to be addressed by a trial were no longer explicitly formulated and expressed in whole sentences.[92] Instead, the minutes, for example of a meeting on 23 February 1973, included a very simple flow chart, outlining which patients should be included in the proposed trial ('patients presenting with disease localized in the chest but not suitable for curative surgery'), and showing a bifurcation point under the label 'Randomization', followed by two columns of text for the 'Delay treatment' and 'Immediate treatment' groups, with different options listed under 'Immediate treatment'. The question (i.e. does treatment with irradiation or cyclosphosphamide, or a combination of the two keep patients alive for longer if it starts immediately than in a control group treated with placebo until they show symptoms?) was implicit in the juxtaposition of the options. The following meeting in July 1973 discussed preliminary results of a chemotherapy trial in Oxford that had started in the 1960s, which appeared to show that there was no difference in terms of survival between the no-treatment group and the group who received treatment (in fact, as it turned out, the patients in the no-treatment group fared better than those receiving chemotherapy).[93] However, while there was recognition that it might be difficult to secure the collaboration of chest physicians in the future, there is little evidence in the minutes of the depressed atmosphere observed at comparable meetings in the 1960s. The working group moved on relatively swiftly to plans for future trials, again presented in the shape of flow charts. The chairman agreed to draft a more detailed protocol and a flow chart, which Fox would use in his discussions with clinicians, and the group agreed to meet again after a few months.

The protocol-driven approach appears to have been fairly well established among members of the lung cancer working party and their collaborators by 1974.[94] The move to the new approach brought with it a much smoother way of working than had been evident during Scadding's chairmanship. While the previous generation had spent much time and energy on discussing fundamentals, in the 1970s and 1980s discussions were mostly about practicalities. The interactions and channels of communication between members of the working party and researchers contributing to trials were more formalized than they had been in the previous decade. Newsletters provided frequent updates. The fundamentals, it seems, were inscribed in the protocols and no longer needed discussion. Nevertheless, the usefulness of chemotherapy for patients suffering from non-small lung cancer remained controversial, and all protocols remained experimental. Chemotherapy was generally reserved for patients with metastatic disease, as a final option to be seen to do something.[95]

One of the medical oncologists who specialized in the treatment of lung cancer in the 1980s was Nick Thatcher at the Christie Hospital. Thatcher had followed Derek Crowther to Manchester, one of the new cohort of research-minded medical oncologists who had trained with Gordon Hamilton-Fairley at St Bartholomew's Hospital in London and were interested in exploring new chemotherapy regimens beyond by then well established applications for leukaemias and lymphomas.[96] Crowther had been appointed to a CRC-funded chair in medical oncology in 1973. As a young doctor at Bart's, Thatcher had initially been working in endocrinology and paediatric oncology. He completed a

PhD at Manchester in 1979 on 'Non specific cell mediated immunity in human malignant disease'. Crowther, who was a member of the new committee chaired by Norman Bleehen and involved in the lung cancer trials initiated in this context, got him interested in lung cancer. A joint appointment was arranged for Thatcher at the Christie and Wythenshawe Hospital, Manchester's thoracic centre, where the treatment of a large proportion of the region's 3,000 or so lung cancer patients per year was co-ordinated. Prior to Thatcher's arrival, there were few links between the two hospitals.[97]

A conference paper presented by Thatcher during a symposium on 'Recent advances in the biology and treatment of solid tumours', sponsored by Lilly Oncology UK and held in Birmingham on 7–8 October 1994 shows that the use of chemotherapy for non-small cell lung cancer remained both experimental and to some degree controversial, but was often demanded by patients who wanted something to be done when other options seemed exhausted.[98] Thatcher presented chemotherapy as an answer to the question of what doctors could offer the 70 to 75 percent of patients whose cancers were not of the small cell type (Non-Small Cell Lung Cancer, NSCLC), but for whom surgery nevertheless did not work: 'A very large number of patients are dying from advanced NSCLC, and more effective chemotherapy offers the only realistic possibility of improving their survival.'[99] Although the amount of time patients gained was fairly limited, Thatcher argued, cancer patients were 'usually much more willing to undergo intensive treatments associated with substantial toxicity for what health professionals may see as minimal or no benefits in terms of the chance of a cure, prolongation of life or symptom relief'.[100] In the discussion following his presentation and a talk by Heine Hansen from Copenhagen, Thatcher argued that he 'would agree that the patient is unlikely to understand median survival improvement, but most patients can understand that with chemotherapy they have a greater chance of surviving 1–2 years than without chemotherapy'.[101] He cited a trial and an MRC meta analysis which had shown that 'the improvement at 1 year with chemotherapy was about 10%. You may deride that figure, but for the patient who is trying to work towards a point in time – an anniversary, a birthday, the birth of a grandchild – even a 10% chance of achieving that end is important'.[102] A straw poll among the audience showed, however, that opinions remained divided. According to the discussion chair, Jim Carmichael: 'Before this debate, out of 52 responses to this question "Is there a role for systemic chemotherapy in NSCLC?", we had 26 in favour of chemotherapy, 23 against, one "don't know", one "maybe" and one "yes and no". A show of hands now suggests the debate has not changed anyone's mind.'[103]

Classification, staging and the emergence of an international organization of lung cancer specialists

The adoption of the protocol approach in trials made their organization simpler and more straightforward and also led to changes in routine treatment. Better outcomes were achieved especially for patients diagnosed with small cell lung cancer, the variant of the disease which responded best to chemotherapy and was most recalcitrant when it came to surgery. A different development, however, was more important to the surgeons involved in the 'normal' treatment of the non-small cell forms of lung cancer, where the benefits of chemotherapy were less obvious.[104] This was the announcement in 1986 and widespread adoption of a new international system of staging lung cancer.[105]

Staging is the procedure of assigning a simple code to each cancer patient, according to an established set of rules, effectively classifying patients with respect to the anatomic extent to which they are affected by the disease. Practitioners have described the accurate staging of lung cancer as 'essential in planning treatment' and 'of crucial prognostic significance'.[106] Staging was also essential when different modalities of treatment were to be compared and results were to be communicated in meaningful terms between different treatment centres.[107] The development of international standards of lung cancer staging, the authors of such statements seem to suggest, was necessary and just a question of time. I argue that it is more useful to interpret the new international staging system as another expression of generational change, as well as successful attempts to establish an international organization dedicated to lung cancer treatment and, not least, the emergence of a secondary specialty among surgeons and physicians treating lung cancer patients that is now known as thoracic oncology.[108]

The staging system was not the first attempt to implement international classifications for cancer, but one more step in a process that started much earlier, driven partly by initiatives of the World Health Organization (WHO) and partly by the International Union against Cancer (Union Internationale Contre le Cancer, UICC). The standardization of disease, along with the standardization of biological substances was among the core activities of the WHO and its precursor organizations. A subcommittee of the WHO Expert Committee on

Health Statistics published a report in 1952, in which general principles of a statistical classification of tumours were discussed. The committee members agreed that three separate classifications were needed, according to (1) anatomical site, (2) histological type and (3) degree of malignancy.[109] A classification according to anatomical site was already available as part of the WHO's International Classification of Diseases.[110] In 1956 the WHO Executive Board passed a resolution asking the organization's Director General to explore the possibility of setting up centres that would arrange for the collection and histological classification of tumour tissues. The resolution was endorsed by the Tenth World Health Assembly in 1957. A study group met in Oslo in the same year to discuss the implementation, and from 1958 onwards the WHO established 23 centres that co-ordinated the histological typing of tumours at various anatomical sites, involving more than 300 pathologists from over 50 countries. The first of these centres, established in 1958, was the International Reference Centre for the Histological Definition and Classification of Lung Tumours directed by Leiv Kreyberg at the Institute for General and Experimental Pathology of the University of Oslo. Based on discussions among an international group of pathologists at meetings in 1958 and 1964, *Histological Typing of Lung Tumours* in 1967 was the first of 25 volumes published over more than a decade and classifying tumours at virtually all known sites.[111]

The WHO's histological typing system matched the second type of classification called for by the organization's Expert Committee on Health Statistics in 1952 and was primarily directed at pathologists. Staging systems, while predominantly anatomical, also responded to the need to classify according to malignancy, in order to help with prognoses. They were addressed primarily to clinicians, both surgeons and physicians. The initiative to establish an international staging system specifically for lung cancer gained momentum in the 1970s and was closely linked to the newly founded International Association for the Study of Lung Cancer (IASLC).[112] However, like the WHO's histological classification, lung cancer staging had its roots in the 1940s and 1950s. It was an application of the Tumour, Node, Metastasis (TNM) system for cancer staging developed for the Enquête Permanente Cancer (French Permanent Cancer Survey) in the 1940s by the surgeon, director of the *Institut Gustav Roussy* and one-time president of the UICC, Pierre Denoix and his colleagues. The TNM system was adopted by the UICC Committee on Tumour Nomenclature and Statistics as the basis for the classification of the anatomical extent of cancer in 1953 and incorporated in the first edition of the UICC manual, *TNM Classification of Malignant Tumours* in 1968.[113] The system was initially criticized as too simplistic but Denoix defended the method worldwide as practicable because of its simplicity.[114] TNM was embraced by the American Joint Committee on Cancer Staging and End Results Reporting (AJC). The AJC was established in 1959, sponsored by the American College of Surgeons, the American College of Radiology, the College of American Pathologists, the American Cancer Society, and the National Cancer Institute, to develop a system of clinical staging of cancer by site. Subcommittees designated as 'task forces' worked on the different anatomical sites.[115]

In a brief presentation to the Sixth National Conference in Denver in 1968, David Carr, Professor of Medicine at the Mayo Clinic Medical School in Rochester, Minnesota, announced that the Task Force on Lung Cancer of the AJC was working on a staging system for lung cancer based on TNM.[116] The other main promoter of this lung cancer classification effort, besides Carr, was the surgeon Clifton Fletcher Mountain, chairman of the Department of Thoracic Surgery at the University of Texas MD Anderson Hospital and Tumor Institute in Houston.[117] The AJC officially published their system of lung cancer staging in 1979. The TNM Committee of the UICC had presented its own lung cancer classification a year earlier, in 1978. Both systems were very similar, almost identical in many points, and plans emerged for a joint publication of a revised system in the 1980s.[118] In 1986 Mountain proposed a new international staging system for lung cancer in the journal *Chest*, which, he thought, resolved the differences between the AJCC (in 1980 the AJC had changed its name to American Joint Committee on Cancer) and UICC classifications.[119] The system had five stages for the classification of six groups of patients with similar prognostic expectations and therapeutic options: 0, I, II, IIIa and IIIb, and IV, based on the position and size of primary tumour and metastases, along with other clinically observable parameters.

What effects did staging have on clinical practice? The adoption of staging systems coincided and was to some degree guided by new diagnostic technologies, for example the introduction in the 1970s and increasingly broad availability in the 1980s of computed tomography (CT) scanners.[120] In this context it is important to know that in the early 1970s the Brompton Hospital, for example, recorded a 20 percent 'open-and-close rate', which meant that in one in five lung cancer cases only the operation revealed that the cancer was inoperable.[121] Essentially these patients were undergoing needless surgery. In another

20 percent of cases the resections were incomplete. Surgeons were selecting patients using what one contemporary, Peter Goldstraw says were very crude parameters. Before CT was available, they looked for mediastinal influence with a barium-swallow (a contrast medium) and a conventional X-ray radiograph. CT changed this. In 1979 when chest CTs were becoming available in the UK, suddenly surgeons had three-dimensional images with better resolution at their disposal. 'Sadly', as Goldstraw remembers, 'we made the mistake of thinking it was all true'.[122] They had seen very little before, using imaging techniques that were very specific but extremely insensitive.[123] 'If a barium-swallow shows you an indentation, you have a load of lymph nodes that big, they're going to be malignant. ... But of course, you're going to miss lots of people with smaller malignant lymph nodes.'[124] CT was just the opposite: incredibly sensitive but very non-specific. 'It showed things; little nodules in the lung; enlarged adrenal glands; it showed loads of lymph nodes, which often times were completely benign.'[125] Until the late 1980s, most of the literature discussed what size of lymph node mattered; what other nodules in the lung were important; what enlargement, what features of the adrenal gland indicated metastatic disease, leading to agonizing debates between surgeons. CT initially drove down the resection rate further than it probably should have done. Patients were being turned down for operations because following a CT surgeons concluded that their cancers were too advanced. Radiologists, too, used to looking at chest X-rays and conventional tomograms, may have initially misinterpreted nodules they saw in CT images as metastases.

While CT changed diagnostic practice in Britain, we need to return to the United States in the early 1970s to sketch another development that guided the introduction of the international staging system. The Mayo clinician David Carr was not only involved with the development of staging, he was also a driving force for the launch of the International Association for the Study of Lung Cancer (IASLC).[126] He introduced the idea of forming an international organization during the First International Workshop for the Therapy of Lung Cancer, held in October 1972 at Airlie House Conference Center outside Washington, D.C., and sponsored by the National Cancer Institute.[127] Carr chaired an organizing committee, which also included Clifton Mountain and which prepared the launch of the association, contacting potential members worldwide. The IASLC was formally launched during the XIth International Cancer Congress sponsored by the UICC in Florence in 1974.[128] After a slow start, the IASLC sponsored its First

World Conference on Lung Cancer in 1978, organized by Mountain and held at Hilton Head, South Carolina. Further World Conferences followed first in two-year, later three-year intervals, attended always by well over 1,000 participants. The newsletter of the Association, edited by Heine Hansen in Copenhagen, turned into a regular journal, *Lung Cancer*, in 1985. It contained a growing number of original research papers. An *IASLC Textbook of Lung Cancer* was first published in 1999. The IASLC provided a home for a new generation of lung cancer specialists, finally severing the link with tuberculosis. It is perhaps indicative of the emergence of this specialty that the new journal of the IASLC, launched in 2006, is the *Journal of Thoracic Oncology*.[129] In Britain, as we have heard, Norman Bleehen, the Chairman of the MRC's Cancer Therapy Committee, was a founding member of the IASLC.

The international staging system started by Carr and Mountain was never a complete product, but rather a continuing project. Initially Carr and Mountain, and later an International Staging Committee of the IASLC continued to improve and refine it. Until its fifth revision, presented by Mountain during a workshop on Intrathoracic Staging sponsored by the IASLC and organized by Goldstraw at the Brompton Hospital in 1996, the system was mostly based on a growing number of cases from Mountain's Houston patient database. There was growing dissatisfaction, however, among chest surgeons about Mountain's exclusive approach. During the IASLC's Ninth World Conference on Lung Cancer in Dublin, in 1998, the IASLC Board decided to commit 5,000 US dollars to turning lung cancer staging into a collaborative project. The International Staging Committee was established, chaired by Goldstraw, whose main task was to collect data from members around the world on lung cancer cases treated by all modalities for further revisions of the system, which the committee would submit to the UICC. The 5,000 dollars were not even sufficient to cover travel expenses to the first meeting, but the members found funds elsewhere and the Cancer Research Campaign (CRC) gave them a committee room and lunch. By 2006, the committee had over 100,000 cases in its database.[130]

Conclusion: Still neglected?

Even after the generational change in the 1970s and despite the international staging system and the rise of the IASLC, which created a new identity for lung cancer specialists, there remained a persistent feeling among those dealing with lung cancer that this was a neglected field. And indeed, as the US-based authors of a comprehensive review article on lung cancer diagnosis and treatment put it in 1986, there was 'controversy on all aspects of the treatment of patients with lung cancer'.[131] Clear-cut reports with reproducible results were the exception, they argued, and accepted standards were lacking. There were only the following 'handful of "axioms" regarding the treatment of patients with lung cancer':[132]

- Where operable, stage I non-small cell cancers should be surgically removed;
- no adjunct therapy had so far been shown to be effective for stage I lung cancer;
- radiotherapy was useful for the palliation of symptoms that could be attributed to local disease; and
- chemotherapy had extended the survival of patients with undifferentiated small cell cancer.

Aside from these 'axioms', therapy was 'largely based on uncontrolled, nonrandomised, retrospective studies' that seemed 'to produce more questions than solutions'.[133]

In Britain, similar sentiments occupied the new generation of oncologists who were specializing in lung cancer, and they looked enviously to what they saw as progress in the treatment of other malignant diseases, especially breast cancer. The Co-ordinating Committee on Cancer Research (CCCR), an umbrella organization for the bodies funding cancer research in the UK, including the Cancer Research Campaign, the Imperial Cancer Research Fund, the Leukaemia Research Fund and the MRC, held a meeting on 3 November 1982 at the MRC headquarters to discuss an initiative to co-ordinate clinical trials in lung cancer, modelled on a subcommittee for breast cancer established in 1978.[134] Some of the statements made in discussions and in the correspondence illustrate my point, and so do the responses to a survey the new lung cancer subcommittee undertook among surgeons, radiotherapists and medical oncologists in 1985. The chairman of the subcommittee, launched in 1984, was the ICRF Professor of medical oncology at Edinburgh, John Smyth, another medical oncologist who had trained with Gordon Hamilton-Fairley at St Bartholomew's Hospital.[135] In a letter to the former ICRF Director of Research and chairman of the CCCR, Michael Stoker, Smyth wrote that, while he was 'enthusiastic about the idea of establishing a subcommittee for lung cancer'; he thought that the principal problem was 'the lack of good quality lung cancer trial work'.[136] Only a small proportion of lung cancer patients, about 2 percent, were entered into any form of clinical trial.[137]

The members of the subcommittee included two further medical oncologists besides the chairman (Nick Thatcher, Manchester and J. N. Whitehouse, Southampton), two radiotherapists (Norman Bleehen, Cambridge and D. Ash, Leeds), two surgeons (Peter Goldstraw, Brompton Hospital and Deirdre Watson, Birmingham), a chest physician (Stephen Spiro, UCL), a pathologist (R. L. Carter, Institute of Cancer Research) and a statistician and epidemiologist (Julian Peto). Smyth appointed the radiotherapist John Yarnold as Scientific Secretary. Representatives of the major cancer research charities, the MRC and the Department of Health and Social Security participated as observers.

Their discussions were frank. This was the commonest cancer, the representative of the Cancer Research Campaign, N. H. Kemp, remarked, and what was needed to overcome the lack of good research was stimulation. The chair admitted that there was 'much pessimism at grassroots level'.[138] Spiro deplored that patients were not entered into good trials, and there was 'a lot of mediocre CT [chemotherapy] given'.[139] The chair suggested that, perhaps, the committee could put 'some moral pressure on people to participate e.g. [in] MRC studies, rather than use protocol + not tell anybody'.[140] What about surgery? The basic questions were answered but more information was needed on 'fine tuning'. Two large MRC surgery trials were about to start. The problem with radiotherapy, according to Bleehen was that many chest physicians did not think of it as of curative or palliative use – 'a hangover from [a] study which wasn't v. good'.[141] The committee decided to send a questionnaire to surgeons, radiotherapists and medical oncologists to find out which trials were going on. The responses confirmed Smyth's observation about the pessimism at grassroots level: One respondent stated, for example, that 'Lung cancer is a waste of time as far as RT [radiotherapy] is concerned'.[142] Another reported:

> Made firm decision a year or two ago to discontinue work in lung cancer field as there is not much interest – presented some important findings at British Association for Cancer Research – to about 5 people in audience. Wonders why this should be so.[143]

In the next chapter I will look at one response to this pessimism at grassroots level: the launch of a charity dedicated specifically to lung cancer in 1990s Liverpool. I will also return to the question of early detection and look at the recent debates around spiral computed tomography as a new technology expected to make screening rational.

7
The Management of Stigma: Lung Cancer and Charity, circa 1990 to 2000

The association with smoking made lung cancer different from other cancers which may have confronted doctors with similar clinical challenges. Lung cancer increasingly came to be perceived as a disease that could not be treated and had to be prevented by persuading people not to take up smoking. This meant that lung cancer sufferers who had never smoked (10 to 15 percent) found themselves associated with a habit which used to be normal but was increasingly seen as a sign of personal weakness and a danger to others.[1] This made lung cancer comparable to tuberculosis and venereal disease, the traditional targets of public health campaigns. The association with smoking led to a common assumption that patients brought the illness upon themselves. But was it ethically acceptable to hold it against smokers that they had conducted a disease that was associated with the habit? How was smoking different from other behaviours which caused illness? In this chapter I will discuss the question of stigma in relation to lung cancer. I will look at the history of the Roy Castle Foundation, a charity dedicated to lung cancer, as a case study of ways of managing such stigma and its consequences. I will also discuss, in this context, some recent developments in lung cancer screening.

Lung cancer and stigma

In ancient Greece, the term stigma referred to a bodily sign that marked a moral fault. A stigma was cut or burnt into a person's skin to label a slave, criminal or traitor, as the sociologist Erving Goffman put it in his much-cited book on *Stigma* in 1968, 'a blemished person, ritually polluted, to be avoided, especially in public places'.[2] More recently, the term has come to refer to conditions that we would prefer to keep hidden from other people because they signalize deviance from dominant norms or are consequences of behaviours considered as immoral: the most obvious are sexually transmitted diseases, the bodily markers of illegitimate sexual contacts. Stigma attached to HIV-AIDS, for example, has been subject to much recent scholarship.[3] Much has been written also, following Goffman's work, on the stigma associated with disability or mental disease. Lung cancer, because of its association with a behaviour whose moral status has changed in recent years, provides a particularly interesting case, combining the stigma traditionally attached to cancer, a deadly disease leading to bodily disintegration, which is said to be increasingly overcome for other malignant diseases, with the stigma associated with conditions that result from habits considered morally deviant. The association with an increasingly 'denormalized' habit fits Goffman's stigma theory well: it affects the social identity of an individual and makes stigma management necessary. Patients diagnosed with lung cancer increasingly face having their identity reduced to one aspect: being a smoker who had it coming or a non-smoker who did not deserve it.

How do we know whether a disease or a habit carries a stigma? We find indications, for example, in recent media coverage. The Scottish edition of the *Sun* newspaper in December 2009 reported on the case of a 'bubbly mum-of-two, from Dundee [who] had always enjoyed robust health – but all that changed with a shock diagnosis in April. Catherine, 60, had lung cancer'.[4] The paper quotes her as reflecting: 'I could hardly say "why me?" – I'd been smoking since I was 15 and we all know the dangers. I smoked my last cigarette the night before I went for surgery to remove part of my lung'.[5] Smokers have long been aware, it seems, that there are risks to health associated with the habit. Many were aware of this by the late 1950s, as we have seen in Chapter 4, and the message has been communicated increasingly forcefully in the following five decades. Knowledge about the risks is also evident, for example, from a comment on the *Daily Mail* website in response to an article reporting the death in 2009 from throat cancer of Alan Landers, a male model known as the Winston Man, who in 1987 had first been diagnosed with lung cancer.[6] Libby from London commented:

> I stopped smoking in December because I had had enough of it, it smells, it's expensive and I was becoming a bit of a social outcast. ...
> I had been smoking properly since I was 15 (27 years) thinking like my other school friends that it was 'cool'. However, I always knew it

was bad for me and my understanding is that we have known for well over 40 years that smoking kills.[7]

Coverage in the press is one indicator, but possibly not a very reliable one. A growing number of expert commentators in recent years, however, both in the UK and elsewhere have suggested that smoking – and lung cancer by association – have indeed become stigmatized. Sociological research undertaken by Alison Chapple and her Oxford colleagues led them to the conclusion that among the 35,000 or so lung cancer patients per year in Britain who do smoke or have smoked in the past, many registered feelings of guilt and shame; some even feeling unworthy of treatment as they thought they had brought this disease upon themselves.[8] In a book on lung cancer for patients and carers, the American radiologist Claudia Henschke and the founder of the Alliance for Lung Cancer Advocacy, Support and Education, Peggy McCarthy, suggested in 2002: 'One of the tragedies of lung cancer is that on top of everything else, patients often feel overwhelmed by guilt. And sometimes they're stung by criticism from family, friends, and even healthcare providers.'[9] These negative feelings, the guilt and the stigma, Henschke and McCarthy argue, 'are not merely sad; they're deadly. In part because of pervasive negative feelings about smokers (and even ex-smokers), many lung cancer patients aren't offered the aggressive treatments routinely provided for those with other types of cancer. And guilt-ridden patients don't demand measures that could prolong and improve their lives.'[10] In a conversation with the author in 2002, Nick Thatcher, a leading medical oncologist in the UK, specializing on lung cancer at Manchester's Christie Hospital, reported similar attitudes in Britain: medical researchers who wanted to undertake clinical trials with lung cancer patients, he suggested, would always have to struggle for resources because cures and treatments for cancers not perceived as self-inflicted were seen as more important.[11]

Is the amount of research funding spent on particular cancers a reliable way of identifying neglect? Also in 2002 the UK National Cancer Research Institute (NCRI) produced what was thought to be the world's first detailed breakdown of cancer research in any country.[12] According to Zosia Kmietowicz in the *British Medical Journal*, the Institute estimated that £250 million from 15 leading charities and government organizations went towards cancer research in the United Kingdom per year. Most of this money was spent on research on the biggest cancer killers. However, some cancers that caused few deaths attracted a disproportionally large share of the funding, while others with a high death toll received rather less attention. The former was especially true for leukaemia (3 percent of cancer deaths and 18 percent of the funding) and the latter for lung cancer (22 percent of deaths, some 3 percent of the funding). The biggest proportion of cancer research spending (41 percent) went on basic biological research, the results of which were assumed to be applicable to all malignant diseases. This was followed by research into treatment (22 percent) and on causes (16 percent). Research into prevention, care for patients and survival attracted the least funding. Similarly, for the United States, the health writer Tara Parker-Pope suggested in her *New York Times* blog *Well* in March 2008 that: 'The big loser in the cancer funding race is lung cancer.'[13] Lung cancer was the biggest killer, she pointed out, but received the least funding, per death, from the National Cancer Institute (NCI). Parker-Pope calculated that the NCI in 2006 spent 1,518 dollars per new case of lung cancer and 1,630 per lung cancer death, compared to 2,596 per new breast cancer case and 13,452 per breast cancer death. Prostate cancer received just 1,318 dollars of research funding per new case, but 11,298 per death. The blog posting was followed by a lively debate with nearly 200 contributions. Several of the correspondents disagreed with the implicit suggestion that more money should be spent on researching lung cancer, a self-inflicted disease: prevention was the key; stop smoking! Others commented on the stigma attached to all cancers, but especially lung cancer.

A few of the respondents to Parker-Pope's blog commented on potential problems that might result from allocating research funding on a per-case, or per-death, basis. Was it not more important to think about what channel of research would provide more new and valuable information per dollar? Neil Burnet, a Cambridge radiotherapist and his colleagues, associated with the University of Cambridge and Addenbrooke's Hospital, proposed more sophisticated ways of assessing research funding per cancer in the *British Journal of Cancer* in 2005.[14] Burnet and his co-authors argued that it made more sense to measure the population burden from different cancers in Years of Life Lost (YLL). They calculated values for YLL for the main cancer sites based on data from the East Anglian Cancer Register for the five-year period from 1990 to 1994, compared with the average life expectancy in the county for the same period. They then plotted YLL for these particular cancers as a percentage of total years lost from all cancers against the percentage of total NCRI research funding spent on these cancers. The results are slightly less bipolar than those compiled by Kmietowicz for the *BMJ*, but they are still striking: Burnet and his colleagues found that lung cancer was responsible for 18 percent of years of life lost due to cancer – by far the biggest killer also by this measure – compared to 3.5 percent of NCRI funding. On the other hand, leukaemia commanded 17.5 percent of the funding but was only responsible for 3 percent of YLL. By this measure, too, lung cancer research, indeed, appeared to be seriously underfunded, vindicating observations such as those by Henschke and McCarthy or Thatcher.[15]

Henschke and McCarthy went further, however, suggesting that lung cancer patients were not only neglected by researchers but also presented with fewer options when it came to therapy. The *British Medical Journal* in 2007 published a debate on 'Should smokers be refused surgery?' between Matthew Peters, associate professor in thoracic medicine in Concord, Australia and Leonard Glantz, a bioethicist based at Boston University School of Public Health. Peters argued yes, smokers who were unwilling or unable to quit, should be refused certain elective surgical procedures where, given all other clinical features were identical, 'costs are increased and outcomes are worse in a smoker than in a current non-smoker'.[16] He specifically mentioned plastic and reconstructive surgery and some orthopaedic procedures, where continued smoking demonstrably increased the risk of complications. Glantz in his response called this position unacceptable and mean. The suggestion that smokers should be deprived of surgery indicated 'that the medical and public health communities have created an underclass of people against whom discrimination is not only tolerated but encouraged'.[17] It is perhaps worth mentioning that most of the correspondents responding to the debate in the journal supported Glantz's position, some arguing that if Peters should be followed, other activities such as certain sports, or conditions such as obesity would also have to be discriminated against. And Peters, too, only argued for restrictions on elective procedures, not life-saving treatments.

Smoking has acquired negative associations in increasingly wider sections of society in recent years, but it certainly has not always been a habit that marked out those indulging in it as less deserving of benefiting from research efforts or treatment. As I pointed out in Chapter 4, the great majority of men in Britain and elsewhere in the industrialized West were smokers when the first results of new epidemiological studies linking cigarettes with lung cancer and other disease were first publicized in the 1950s. Rosemary Elliot has analysed 126 interviews with ordinary Britons born between 1906 and 1972 and found that taking up smoking was then a normal rite of passage, marking the beginning of adulthood.[18] However, when the interviews were recorded in 1986, smoking clearly had already acquired predominantly negative connotations and was associated with ill health. The American economist W. Kip Viscusi studied the perception of risks associated with smoking in a sample of the US population in the early 1990s. He found, perhaps surprisingly, that respondents overestimated rather than underestimated health risks. When asked how many out of 100 smokers they thought would die from lung cancer, heart disease, throat cancer or any other illness caused by smoking, the average estimate was 54 – two to three times as high as the actual risk of dying from smoking-related illnesses (the actual figure is between 18 and 36). Viscusi's respondents assumed that 38 out of 100 smokers would die from lung cancer alone, which is four to eight times as many as the actual figure (the scientific consensus in the 1990s was that five to ten out of 100 life-long smokers faced a lung cancer diagnosis). Interestingly, smokers as well as non-smokers overestimated the risks.[19]

If even exaggerated assumptions about the health risks resulting from smoking have not persuaded a great number of smokers from indulging in the habit, were there other factors? In most cases, smokers started as children.[20] There was obviously the controversial matter of addiction – and addictions often carry stigma – but also the assumption that certain personal characteristics – such as neuroticism – were associated with the habit and the inability to quit.[21] The link between personality and smoking (and personality and lung cancer) has long been a subject of research and debate. The maverick psychologist and professor of psychology at the University of London, Hans J. Eysenck was one of the prominent critics of the smoking-causes-lung cancer hypothesis in the 1950s, along with, for example, the statistician R. A. Fisher.[22] Following Fisher, Eysenck argued that both the rate of smoking and the proneness to lung cancer could be attributed to a third variable underlying both. Fisher proposed that there was a genetic factor. Eysenck used psychological research methods to test the hypotheses that cigarette smokers were (1) more extraverted, (2) less rigid, and (3) more neurotic than non-smokers. He and his co-authors strongly confirmed the first hypothesis, weakly confirmed the second, and failed to confirm the third. They also found that pipe smokers were introverted and concluded that 'genotypic differences exist between smokers and non-smokers, and between cigarette smokers and pipe smokers'.[23] Following this study, Eysenck began to collaborate with David Kissen, a Glasgow-based chest physician interested in psychosomatic illness, who had developed similar theories about the origins of lung cancer.[24] Kissen had become interested in emotional factors in pulmonary tuberculosis as a medical officer in a mental hospital and a sanatorium, and as area chest physician in Lanarkshire.[25] In 1958 he had joined the staff of the Southern General Hospital, Glasgow and extended his research on personality factors to lung cancer and smoking, presumably partly as a consequence of the broad interest in these issues at the time. This work gained him research grants, allowing him to set up a psychosomatic research unit at the hospital. He carried out extensive interviews with 900 patients in a number of chest departments, whose diagnoses were unknown to both him and the patients themselves at the time of the interview. About half of these patients were subsequently found to suffer from lung cancer, the other half served as controls. Based on these interviews, Kissen concluded that lung cancer patients appeared to have 'poor outlets for emotional discharge', no matter whether they smoked or not.[26] Kissen provided Eysenck with empirical data and Eysenck presented Kissen with a classification scheme with quantifiable personality measures – Eysenck's so-called Maudsley Personality Inventory (MPI) – which enabled Kissen to substantiate his essentially clinical conclusions. Low scores on the 'neuroticism scale' of the MPI represented an individual's emotional instability, over-responsiveness and liability to neurotic breakdown, and thus, to Kissen, suggested an inability to discharge emotions, an unconscious repression and denial of emotional difficulties. Smokers in general scored high on extraversion and fairly average on neuroticism, suggesting to Kissen and Eysenck that lung cancer patients were not a random sample of cigarette smokers, but a special group, and that the same applied to non-smokers who developed lung cancer.

Kissen's and Eysenck's research in the 1960s left lung cancer sufferers with a stigma of sorts – what do you do about an inability to discharge emotions? – but not smokers in general, whose personality attributes according to MPI were fairly attractive. However, it became increasingly clear that there were links between social class and smoking habits, leading to another potential source of stigma. Since the 1950s the difference between the sexes as regards smoking habits became smaller while class differences grew.[27] Doctors and other professionals were the first to stop smoking (see Chapter 4). The statistician G. F. Todd reported in 1976 that among professional men, the average weekly consumption of cigarettes had declined from 76 in 1958 to 44 in 1974.[28] In contrast, while an unskilled worker had smoked in average 78 cigarettes per week in 1958, he was smoking 99 per week in 1974. The average for men in all social classes was 76 in 1958 and 74 in 1974. Over the same period, average cigarette consumption among women in all social classes apart from professionals went up from 31 per week in 1958 to 50 in 1974. These developments had their echo in lung cancer incidence figures with about two decades delay. Todd found that during the time period he studied 'the social class gradient of lung cancer in men increased substantially' and 'a smaller social class gradient' developed in lung cancer rates among women.[29] The trend did reverse, however. In the 1980s and 1990s, men and women in England in the socioeconomic groups classed as manual workers followed those classed as non-manual in reducing their cigarette consumption.[30] It seems that the anti-smoking campaigns had some success after all: people were quitting. Smoking, in turn, increasingly signalled membership of the lower social classes.

But there was more to it than simply class difference: smoking was also increasingly associated with delinquency. An article in the online version of the *Sun* newspaper in 2006 reported results of a study undertaken by researchers at Bournemouth University under the headline 'Smokers' kids "are yobs"'.[31] The Professor of Psychiatric Social Work at Bournemouth University, Colin Pritchard and colleagues had found that children of smokers were four times more likely to smoke themselves, twice as likely to steal, get into fights and become sexually active at an early age, two and a half times more likely to take drugs or binge drink and four times more likely to have unprotected sex than children of non-smokers.[32] Let's emphasize: we are not talking about the children themselves being smokers, but their parents! The association of smoking with delinquency, however, was not new: a study undertaken by J. W. Palmer at the MRC Epidemiological Research Unit in Cardiff, for example, found in 1965 that in a secondary modern school where caning was used as punishment for smoking, caned boys were considerably heavier smokers than those boys who did not receive the punishment.[33] Twenty-six percent of the caned smokers studied by Palmer had been classified as delinquent in police records; they had committed such offences as larceny (like robbery, but not involving violence, 18.5 percent), trespass on the railway (i.e. travelling without ticket, 6.2 percent), breaking and entering (5.5 percent), or malicious damage (5.5 percent). Boys classified as delinquent were significantly heavier smokers than non-delinquent boys, but the difference was less marked than that between caned and uncaned boys. Interestingly, a number of boys referred, unprompted, to the risk of lung cancer associated with smoking during their interviews. As

Rosemary Elliot has argued, there was a difference in the 1950s and 1960s between children who smoked (not acceptable) and adults (respectable). Taking up smoking was widely regarded as a rite of passage marking the transition into adulthood, when it was seen as a legitimate habit.[34] Pritchard and his co-authors in 2006, in contrast, appear to interpret parents' smoking as not only a marker of deprivation, but to some degree also as a cause of delinquent behaviour.[35]

What has been described as the 'denormalization' of smoking started in the late 1960s when a new style of anti-tobacco activism – in Britain epitomized by the campaigns of the group Action on Smoking and Health (ASH, see Chapter 6) – replaced the more traditional approach aimed at education and persuasion.[36] The new campaigns aimed to change the image of smoking, making it socially unacceptable and turning smokers into outcasts – quite successfully as not only the quotes from the tabloid press suggest. Smoking was gradually reframed as an environmental health issue, and second-hand smoke came to mark a particular battle ground, based on scientific evidence which was far from definitive. Social norms were changing. In an article published in 1979 the American sociologists Gerald Markle and Ronald Troyer compared the anti-tobacco with the temperance movement and wrote that, by the late 1970s: 'In addition to being seen as harmful to health, smoking came to be seen as undesirable, deviant behaviour and smokers as social misfits. In fact data shows that people increasingly view smoking as reprehensible.'[37] Christopher Snowdon in his book on the history of anti-smoking argues that since the 1980s a growing number of articles aiming to prove health risks associated with passive smoking based on shaky evidence were published by medical journals because they supported what was now considered a good cause.[38] The *Journal of the American Medical Association* under the editorship of George Lundberg published a number of studies on passive smoking which Snowdon calls questionable.[39] An article published in the journal in 1998, for example, appeared to demonstrate that individuals exposed to second-hand smoke had a higher risk of hearing problems than smokers, a result that is rather implausible if the hearing loss is caused by tobacco exposure.[40] Methodologically sound epidemiological studies, on the other hand, whose authors concluded that increased risks to non-smokers exposed to second-hand smoke could not be demonstrated were exposed to attacks and their authors (along with the responsible journal editors) denounced as stooges of the tobacco industry.[41] Snowdon quotes Richard Doll who told a radio station that 'the effect of other people smoking in my presence is so small it doesn't worry me'. Doll, who can hardly be accused of being a puppet of the tobacco industry, found himself a target of outraged critics, compelling him to point out that he was only 'speaking personally'.[42] The ultimate goal was a ban of smoking in public places, as pioneered, for example in California and as implemented in Scotland in 2006, and England, Wales and Northern Ireland in 2007, along with a further marginalization of smokers.[43]

This approach may not necessarily be productive. Scholars studying its effects in the context of the AIDS epidemic found that stigmatization and marginalization subverted prevention efforts. Ronald Bayer, Professor at the Center for the History and Ethics of Public Health in the Department of Sociomedical Sciences at the Columbia University Mailman School of Public Health, suggested in 2006 that there were obvious comparisons between HIV-AIDS and tobacco control.[44] Bayer and his co-author Jennifer Stuber argued that anti-smoking activists rarely addressed 'the moral question of how to balance the overall public health benefit that may be achieved by stigmatization against the suffering experienced by those who are tainted by "spoiled identities"'.[45] This was all the more pressing as stigmatization fell on the most socially vulnerable: the poor who continued to smoke. In the remainder of this chapter I will use the history of a charity founded in the 1990s in Liverpool as a case study of how the stigma of lung cancer was addressed in a city that was plagued by a range of social problems.

Charities and the de-stigmatization of cancer

Charities have long played an important role when it came to funding cancer research and cancer hospitals. Charities, as Emm Barnes shows, contributed to changing the face of childhood cancer since the 1960s, turning their campaigns into models for the fundraising activities of other cancer charities.[46] Charities also contributed to de-stigmatizing cancer. Since the 1990s, charity campaigns have endorsed a new 'culture of survivorship', with patients no longer shying away from reporting publicly on their experiences with illness and treatment. This has been especially notable for breast cancer in America, so much so that critics have spoken of a 'tyranny of cheerfulness'.[47] In the UK, breast cancer has become similarly de-stigmatized, with super models and other celebrities publicly supporting breast cancer appeals.[48] Breast cancer, of course, never carried the moral stigma that lung cancer has acquired since the 1950s. Moreover, lung cancer neither possesses the sex appeal of the recent breast cancer campaigns nor does it generate images as iconic as the bald children undergoing cancer chemotherapy often employed by campaigns against leukaemia and childhood cancer. Until fairly recently, there were no charities dedicated to lung cancer treatment or research, neither in the UK nor the US.

As Ilana Löwy, Jean-Paul Gaudillière and others have shown, much basic research in cell and molecular biology has been labelled cancer research.[49] Such research has often been funded by large and influential cancer charities, such as in Britain the Imperial Cancer Research Fund or the Cancer Research Campaign, the precursor organizations of today's Cancer Research UK, and in America the American Cancer Society.[50] Many of the researchers pursuing projects funded as cancer research, however, as Gaudillière's work shows, only had a marginal interest in the disease; their focus was the immediate object of their work: for example a virus, an innovative technology, a gene or protein molecule.[51] Critics have pointed out that with a view to actual therapeutic progress, the unprecedented investments in fundamental cancer research since the end of the Second World War have delivered disappointing results.[52]

Fundamental research, however, is only one way in which modern societies have confronted cancer. There are other ways of dealing with the disease. Therapeutic innovations were often products of clinical research and local tinkering. Moreover, despite – or maybe because of – the dreadful connotations that the disease carries, specialist cancer hospitals, at least in Britain, tend to be places with which local people have long identified, almost like churches.[53] People wanted to be proud of their local cancer hospital, a place where heroic high-tech battles were fought against an invisible enemy. But charities have left their marks not only where it came to supporting fundamental research or local cancer hospitals. Thanks to hospice charities, Macmillan Cancer Care, Marie Curie Cancer Care and similar organizations, there are fewer messy or lonely cancer deaths today in dirty homes or dark hospital side wards (see also Chapter 6).[54] More so than is the case for any other disease, cancer deaths in the UK are increasingly well managed; they have become meaningful and invested with positive connotations. I will now turn to the history of the UK's first specialist lung cancer charity, the Roy Castle Foundation.[55] It was launched in the early 1990s by a Liverpool chest surgeon, Ray Donnelly, to fund basic research on lung cancer, as he felt this disease was neglected by researchers, other cancer charities and state agencies due to the stigma attached to the disease.[56] The Foundation explicitly embraced the survivorship ethos of recent breast cancer campaigns, aiming at reducing the stigma associated with lung cancer. I will argue, however, that the history of the Roy Castle Foundation is characterized by strong continuities with older traditions of campaigning in Britain – not least by a strong religious dimension – and by significant local specificities.[57]

Faith, place and patrons

Raymond Donnelly, a native Glaswegian, Jesuit-educated and a committed Catholic, came to Liverpool in 1975 as a consultant specializing in pediatric cardiac and adult thoracic surgery. Born in 1936, Donnelly had graduated from St Mary's Hospital Medical School in London in 1961. A Fellow of the Royal College of Surgeons of Edinburgh since 1969, he went to Harvard University as a research fellow working in cardiac surgery in 1973. Towards the late 1970s, after a few years in Liverpool, he increasingly felt that he could no longer do justice to both, operating children at the children's hospital at Alder Hey and adults at Broadgreen Hospital, several miles away. He devoted himself exclusively to adult thoracic surgery from 1979. The Unit at Broadgreen, established by Hugh Reid and Hugh Morriston Davies as part of the Emergency Medical Service during the Second World War (see Chapter 3), was the regional referral unit for thoracic surgery cases not only for Merseyside but also for North Wales and the Isle of Man. Donnelly saw between five and ten new lung cancer patients per week, making it clear to him 'how much lung cancer there was in the region'.[58] He became increasingly frustrated by the low success rate and by how little this had changed since the 1950s: 'The situation was unacceptable. A concentrated effort was required.'[59] Donnelly's answer was to launch a charity dedicated to fundamental research on lung cancer.

The Lung Cancer Fund was born on 18 April 1990 in Donnelly's office at the Cardiothoracic Centre.[60] The surgeon launched the charity with his secretary Sheila Christian and one of his patients, Eric Morris, a local businessman who had had one of his lungs removed for cancer by Donnelly two years earlier. Donnelly's decision to found a charity himself rather than rely on existing organizations was triggered, it appears, in the last instance by a failed funding application to the British Lung Foundation (BLF) and, more broadly, by the impression that the existing cancer charities were not interested in lung cancer. Donnelly was not happy with the dominant notion that all that needed to be done to tackle lung cancer was to stop people smoking. What about the large group of people who heeded the warnings and gave up the habit but nevertheless developed lung cancer? What about those who were diagnosed with lung cancer but had never smoked? He felt that a research programme into the molecular biology of lung cancer was needed to investigate why, when and how some – smokers or non-smokers – got lung cancer and others did not. What were the early molecular and cellular changes that occurred during the course of the disease? Was there a way of diagnosing the disease much earlier, resulting in better chances of a cure? Donnelly remembers that a conference in Toronto in 1986 made him realize the importance of combining molecular biology with epidemiological approaches. With his application to the BLF Donnelly wanted to win funding for a Senior Lecturer post at the University of Liverpool for a thoracic surgeon with a special interest in lung cancer. The post had been approved by the university, but there was no money for it. Donnelly expected that such a post would have stimulated an intensive lung cancer research programme. The BLF turned down the application and there were no other obvious sources of funding. Was the launch of a charity a solution? How did one go about launching a lung cancer charity?

During the meeting on 18 April 1990, Donnelly was elected Chairman, Eric Morris appointed as treasurer and Sheila Christian, who had some experience working with a pain relief charity, became the Lung Cancer Fund's secretary. They decided on the name and a logo, but there was no money in the bank. Less than five years on, by the end of 1994, the charity had collected more than two and a half million pounds for a research centre in Liverpool, which was being built on a plot of land donated by the city council and which opened its doors in 1997. The name of the Fund was changed in 1995 to include the name of the entertainer Roy Castle, who died in 1994 from small cell lung cancer and was the best known patron of the charity. The rapid growth of the Lung Cancer Fund had a lot to do with its patrons. The Fund also had strong local Liverpool roots; and religious faith was central, as we will see. In the following paragraphs I will look at this success story. I will then assess if research funded by this charity was different from science financed by government agencies.

In the first three years of its history, the Lung Cancer Fund focused its fundraising activities almost exclusively on Liverpool and Merseyside. Local newspapers, the *Liverpool Echo* and the *Daily Post*, along with local radio stations and the Liverpool studios of the BBC and the private ITV network provided the publicity. Local businessmen (some of them Donnelly's patients) served as trustees. Local celebrities such as the comedian Ken Dodd or Libor Pecek, principal conductor of the Liverpool Philharmonic Orchestra, volunteered their time and acted as patrons. An important patron in these early years, and the charity's first President, was Lady Mavis Pilkington, the wealthy widow of a St Helens-based glass manufacturer and affectionately known as the Queen of Merseyside. Lady Pilkington gave generously, both time and money. Publicity was also generated by Donnelly himself, who, in cooperation with manufacturers, had for some years developed instruments for keyhole surgery and techniques for chest operations using these instruments. In 1991 he invited the press for a new operation, a world first, using a novel staple gun to remove a lung tumour. The operation on 3 October was covered by television teams and written about in local and national newspapers.[61] Also in 1991, Donnelly's charity began a productive collaboration with Liverpool City Council, developing local anti-smoking campaigns aimed especially at children.

Littlewood's Pools, another Liverpool institution was an important source of support for the Lung Cancer Fund. Littlewood's Pools was a Liverpool company which also ran a well-known mail order business and a chain of department stores. The Moores family, owners of Littlewood's, were important local philanthropists; John Moores senior, the founder of the company funded, for example, the biennial John Moores Liverpool Art Exhibitions since 1957, a school of business and management studies at Liverpool University in 1963, and in 1960 he became chairman of Everton Football Club, attending almost every match. Littlewood's Pools sales were affected by the new National Lottery and they were launching a new scratch card lottery in 1994, which was meant to benefit selected good causes, for which customers could vote. Not least due to personal contacts, the Lung Cancer Fund was chosen as one of the candidates. Littlewood's also provided the Fund with free office space and occasional access to a private jet.[62] John Moores, the son of the Littlewood's founder succeeded Lady Pilkington as president.

It should be clear by now that the Lung Cancer Fund had strong local roots in Liverpool and the Merseyside region, and this is important. Looking back at a proud history as 'the second city of empire', by the 1980s and 1990s the city was viewed as the metaphorical 'sick-man' of British cities by many in Britain, a 'Cinderella city', as it were, and the notion that its people were particularly badly affected by this disease resonated with the narrative of deprivation by then associated with Liverpool. Glasgow, of course, Donnelly's home town and the city where the Foundation opened its second base in 1996, shared this fate and has similarly strong Catholic traditions. The enthusiasm with which both the Merseyside elites and the population more broadly embraced the campaign initiated by Donnelly resonated with another narrative, that of fierce local pride and resilience. Scousers, as Liverpudlians are known colloquially, did not give up easily and stood together when faced with hardship.[63] This is epitomized by the unofficial anthem of the city (and of Liverpool Football Club), the 1940s show tune 'You'll never walk alone', made popular by the Merseybeat band Gerry & The Pacemakers in 1963.[64]

For Liverpool with its magnificent buildings, built on profits from shipping and the trade with sugar, slaves and cotton in previous centuries, the decades from the 1960s to the 1990s were characterized by industrial decline, economic depression, extremely high unemployment (up to 50 percent in some areas of the city) and attempts – often failing – of urban regeneration.[65] The rest of Britain associated with Liverpool the notorious Toxteth race riots, a militantly left-wing Labour council that declared war on the Tory government, and a recalcitrant labour force. The London-based tabloid press promoted an image of the stereotypical Scouser as workshy and involved in petty crime.[66] But there were also, of course, the Beatles and other Merseybeat bands, and two well-known football clubs, Everton and Liverpool. Football was (and is) an important source of the Liverpool identity. One of the key moments of Roy Castle's campaign for the Lung Cancer Fund in 1993 was his appearance at the derby match between Liverpool and Everton at Liverpool's Anfield stadium, where he was invited to toss the coin before the start. The fans sang 'You'll never walk alone' and 'There's only one Roy Castle'. Ray Donnelly remembers that he was 'in [his] usual place in the Paddock and found the whole thing very emotional', and a bucket collection at the gates raised a total of £10,000.[67]

The Roman Catholic Archbishop of Liverpool, Derek Worlock was a patient of Donnelly's and the surgeon managed to recruit him as an early patron of the Fund, illustrating the importance of both the local and the spiritual dimensions to the story of this charity. The London-born Worlock was a well-known man in the city, an important unifier, along with his Anglican counterpart, the popular former England test cricketer Bishop David Sheppard.[68] Worlock and Sheppard played significant roles for consolation and reconciliation after the tragic events at Heysel Stadium in Brussels in 1985 and at Hillsborough Stadium in Sheffield in 1989, which both involved Liverpool football fans. According to the author of Worlock's obituary in the *Independent* newspaper, the two became inseparable and were known as

'Tweedledum and Tweedledee' or 'Fish 'n' Chips' (because they were always seen together in the newspaper), and their partnership helped defuse latent sectarian conflicts between Protestants and Catholics in the city. When he had opened the Archbishop's thorax, Donnelly remembers that '[a]s I held his lung in my hand I had a real sense of spiritual awareness, of responsibility for the life and health of one of the senior prelates of the church who was held in such high regard by people of all denominations ... I prayed that God would guide my hand'.[69] The operation was successful and was followed by chemotherapy. During Worlock's stay in hospital after the operation, Donnelly talked to him about the Lung Cancer Fund and asked him to become a patron. He also suggested that the Archbishop 'should speak out publicly about lung cancer and be a voice for all the thousands of people in the region who developed the disease but did not have the same opportunity as him to be heard'.[70] This appealed to the Archbishop, Donnelly remembers, and he accepted the role. Football and religion were two powerful social forces that Donnelly managed to mobilize for his cause.

Worlock was 72 when Donnelly treated him, a typical age for a lung cancer diagnosis, and he had been a smoker earlier in his life, although he had not smoked for three decades. Another special patient was Nicola Lawrence, a very untypical lung cancer patient. A flight attendant working for the Scottish airline Loganair, she was only 24 years old when she noticed strange and persistent pain in her back and shoulder and found that she coughed up blood during a holiday. The diagnosis took some time because her case was so unusual. When Lawrence was referred to Donnelly for treatment, he told her about the Fund and she started to campaign as much as her illness (and the treatment) allowed her. A young air hostess who had never smoked was a more effective face for a campaign than, to use the words of the chest surgeon Sir Thomas Holmes Sellors, the typical 'emphysematous and bronchitic elderly men who smoke heavily'.[71] As is the case for many young patients, Lawrence's cancer was particularly aggressive; it was no longer operable and she died only four months after she had been referred to Donnelly, in February 1993. Her friends and colleagues continued to raise donations for the Nicola Lawrence appeal, totalling £100,000. The money was used to build and equip the library of the lung cancer research centre. The library was named after her and a photograph of the flight attendant in her uniform displayed outside the door as a memorial.

This personal dimension may appear overly sentimental to some readers, but it is not unusual for cancer charities. Appeals are often personalized and laboratories, buildings, or rooms named after patrons who frequently were patients. Other examples include Dimbleby Cancer Care, a charity founded in the memory of Richard Dimbleby, the well-known broadcast journalist who died from cancer in 1965 and, unusual by the time, went public with his diagnosis, or Maggie's Cancer Care Centres, launched by and run in memory of Maggie Keswick Jencks, a writer, artist and garden designer who died from breast cancer in 1995.[72] Religion is another common feature: according to the author of her obituary in the *Independent* newspaper, the Roman Catholicism in which Jencks was raised remained a strong influence during her life and beyond her death. The spiritual dimension, of course, is also central to the hospice movement.[73]

Donnelly credits Nicola Lawrence with persuading the perhaps most important patron and future figure head of the new charity to take an interest in the Lung Cancer Fund, the popular entertainer Roy Castle. Castle was a trumpeter, singer and dancer who held the world record for the fastest tap dance. He was praised for his 'happy-go-lucky charm and the sheer freshness of his performance' and was often called the British Sammy Davis Jnr.[74] Castle was probably best known by the early 1990s as presenter of the television programme *Record Breakers*, a TV adaptation of the idea behind the Guinness Book of Records, for which he performed numerous dare-devil stunts in order to establish new records. He was a frequent guest on the BBC children's pro- gramme *Blue Peter* and also presented fitness programmes. According to his obituary in the *Guardian* newspaper, 'television was an ideal medium for his homely, modest, yet perky personality'.[75] In March 1992 the entertainer was diagnosed with lung cancer after undergoing extensive medical tests for what he thought was an attack of migraine in January, followed by what he describes in his memoirs as an increas- ingly distressing, claustrophobic feeling around his chest: 'I felt as if an invisible boa constrictor was slowly tightening its grip and trying to suffocate me.'[76] A Computed Tomography (CT) scan revealed a tumour in a position that made it inaccessible for surgery, and the examination of a biopsy sample taken during a bronchoscopy confirmed that the cancer was of the particularly aggressive and essentially inoperable small (or oat) cell type. A life-long non-smoker, Castle blamed passive smoking for his condition: he performed in too many smoke-filled jazz clubs. His prognosis was not good, the consultants told him: without treatment he could expect about another three months of deteriorating health. They recommended a course of chemotherapy, which would give him a five to ten percent chance of long-term remission. He decided to go for the chemotherapy, which started two days after the diagnosis. The chemotherapy was initially successful, but the cancer recurred in 1993.

Castle and his wife of 30 years, Fiona, were both 'strong Christians with a total trust in God'.[77] Donnelly comments on Castle's 'cheerful and uncomplicated acceptance of God's will for him'.[78] The entertainer did not want to die, but if that was what God wanted for him, it was all right by him. According to Donnelly the matter was clear to Castle: 'You get on the bus and you trust the driver.'[79] In the event Castle got onto a train for the Lung Cancer Fund, a vintage Pullman train, in July 1994. It had been offered to the charity for use during what became the 'Roy Castle Tour of Hope' by the Director of Special Trains for British Rail, who was from Liverpool, had been diagnosed with cancer himself a few years earlier, and was an admirer of Roy Castle.[80] As was the case for Nicola Lawrence, the intense campaigning – along with writing his autobiography – appears to have given focus and meaning to Castle's life when he knew that his cancer was terminal. Castle was an ideal figure head for a charity in the age of electronic mass media and consumerism. The Tour went up to Scotland and finished in London, visiting all major cities in the United Kingdom on the way. It was covered by the national and the local press and a TV documentary was recorded. Donnelly counted a total of 119 television slots and 153 radio slots devoted to the Roy Castle Tour of Hope, and by the time the train reached London, £1.3 million had been donated or pledged.[81] Castle was from Yorkshire rather than Merseyside, but according to Donnelly, there was 'a deep bond of affection between Roy and the people of Liverpool'.[82] While the local basis remained crucial, the Roy Castle Cause for Hope Appeal took the campaign beyond Merseyside. Castle died in September 1994. In the following December, he was elected Personality of the Year by the listeners of BBC Radio 4, a station broadcasting spoken-word programmes to an educated and predominantly middle class audience.

The early campaigns of the Lung Cancer Fund and later the Roy Castle Foundation, at first glance, were above all attempts to rehabilitate lung cancer as a worthy cause, aiming to remove the stigma of the self-inflicted disease by drawing attention to the victims of passive smoking, the ex-smokers and those 10 to 15 percent of lung cancer patients who had never smoked. The charity also addressed the spectre of hopelessness that stalked lung cancer sufferers and the practitioners treating them by including personal accounts by survivors (preferably young and attractive ones) in their campaign literature. There has been another, more subtle and probably unintended dimension to these campaigns: they also addressed the stigma of hopelessness associated with Liverpool that had its origins in the city's recent history and in southern attitudes towards the industrial North West. The slogan chosen for the first big poster campaign around Merseyside in 1992 was 'Cause for Hope'.

Charity and science

I have argued that the short history of the Roy Castle Foundation has been centrally shaped by patrons and their experiences with lung cancer, by the city of Liverpool, by biographies and personal contacts, and by mobilizing forces such as football and religious faith. I will now attempt to explore if and how this cosmology shaped the activities funded by the charity, especially in the field of medical research.

The Roy Castle Lung Cancer Foundation and similar charities distinguish themselves from government-funding bodies such as, say, the Medical Research Council not only by way of their local roots and the importance of personal contacts between officers (including medical practitioners) and patients and their relatives (including many trustees and patrons). Another distinguishing factor, which these charities have in common with some Victorian hospital charities, such as, for example, the Brompton Hospital for Diseases of the Chest, is the focus on one particular disease or group of diseases.[83] The motivation for launching the Lung Cancer Fund grew out of Donnelly's daily experiences interacting with lung cancer patients and the neglect and disinterest that he believed kept this disease from receiving the attention and the research funding it deserved considering its apparent significance for public health. The motivations of many of the patrons, trustees and supporters originated from experiences with lung cancer in their circles of friends or relatives. These people were not particularly interested in fundamental research that explored what went wrong in cells when they transformed into cancer cells. They wanted to see research being done that directly tackled lung cancer, the illness they knew, and which promised to make a difference if not for today's sufferers then for future patients diagnosed with this particular malignancy.

The cosmology that charities such as the Lung Cancer Fund embody lead many of them not to fund research but care centres and other measures that directly improve the lives of patients once they have been diagnosed with cancer. Typical examples are Dimbleby Cancer Care or Maggie's. They aim to provide dignified surroundings and support to often terminally ill patients, in the spirit of Cicely Saunders' hospice movement, Macmillan Cancer Relief or Marie Curie Cancer Care.[84] In fact, the first measures funded by the Lung Cancer Fund were a specialist lung cancer support nurse and a pleasantly furnished counselling room at the Cardiothoracic Centre where Donnelly worked, sponsored by (and named after) one of the early patrons of the charity, where newly diagnosed patients could talk to the nurse over a cup of tea about what the diagnosis meant for them.[85] The foundation later funded lung cancer nurses in many other hospitals, whose salaries usually were covered after a while by local NHS trusts. The foundation also sponsored patient support groups, especially from its second base in Glasgow. One of Donnelly's patients, Terry Kavanagh, a Liverpool joiner, life-long non-smoker and small cell lung cancer survivor of more than 15 years, who raised more than £200,000 for the foundation, has also become a leading cancer patient advocate and the UK's lung cancer representative on the Scientist/Survivor Programme run by the American Association for Cancer Research.[86] However important these activities are for the public face of the charity and its appeal to donors, Donnelly explicitly wanted the organization to sponsor basic research that nobody else would fund, an 'intensive laboratory research programme'.[87] His ultimate aim was a purpose-built research centre dedicated to lung cancer.

In 1993, after Donnelly had started to put out feelers around Liverpool University he was approached by a geneticist who had been working in the University's dental school for some years on the genetic characteristics of head and mouth cancers, Dr John Field. Field developed a proposal to study early genetic changes occurring in lung tissue during the development of lung cancer. The application was reviewed by specialists outside Liverpool and submitted to the Trustees. At this point there was not enough money in the Foundation's coffers, but the contact was established and Field became its Director of Research, with an annual research budget, in 2006, approaching £1 million.[88] Donnelly and the trustees 'held the firm view that early detection was the future so far as lung cancer management was concerned'.[89] The first goal of what came to be known as the 'Liverpool Lung Project' was to develop means of detecting subtle genetic changes in the cells of the lung epithelium – the lining covering the inside of the lungs – before they had completed the transformation into cancer cells, reminiscent more of the precancerous lesions identified in cervical cancer screening programmes than the breast cancer genes BRCA1 and BRCA2. The next step was to develop non-toxic agents that arrested the process of transformation and prevented the development of a malignant tumour. The models here were breast cancer and tamoxifen, and the ultimate goal the effective chemoprevention of lung cancer.[90]

The idea of detecting lung cancer early and thus improving the chances of treating it successfully was not new, as we have seen in the previous chapter. However, as I discussed in Chapters 5 and 6, trials since the 1950s using radiological techniques and the cytological examination of sputum samples had repeatedly shown that detecting tumours in the lung earlier made no or very little difference for survival. Promoters of screening usually responded by arguing that while the principle was good, the technology was not ready. More recently a controversial debate has been unfolding, predominantly in the US, over the use of a new technology: low dose, spiral CT scans for the early detection of lung cancer in patients at high risk of developing the disease (mostly middle-aged smokers).[91] Field and his Liverpool colleagues announced a collaboration with the US researchers involved with the Early Lung Cancer Action Project (ELCAP) in 2001, and more concrete goals were laid down in a 2005 'Liverpool Statement'.[92]

Claudia Henschke and David Yankelevitz, two radiologists and their colleagues based at Weill Medical College of Cornell University and New York Presbyterian Hospital, initiated ELCAP in 1992 to assess the usefulness of annual CT screening for lung cancer. They had developed doubts about the results of the Mayo Lung Project, the main source of evidence against the usefulness of radiographic screening and were encouraged by the greater potential of computed tomography, a more sensitive technique which allowed the more reliable detection of smaller tumours in awkward locations and thus promised to help overcome some of the problems associated with traditional chest X-rays (see Chapter 6). They published their baseline results in the *Lancet* in 1999 and a big article reporting positive results in the *New England Journal of Medicine* in 2006, where they concluded that annual spiral CT screening 'can detect lung cancer that is curable'.[93] The study was praised because it involved so many participants (31,000 in seven countries) and immediately criticized because there was no control group.[94] The scans revealed about 4,000 suspicious nodules; biopsies were taken in these cases, leading to the detection of more than 400 tumours. Critics pointed to the risks associated with a needle biopsy of lung tissue; the CT scans would draw attention to many suspicious-looking nodules which would ultimately prove harmless. They also took issue with the cost of the procedure (Yale University Hospital, for example, charged about $800 for a scan and its interpretation); and drew attention to the paradox that blights all screening programmes: picking up a cancer earlier may prolong the time between detection and death of a patient, but not necessarily the patient's life.[95] The critics pointed to a randomized controlled study organized by the National Cancer Institute, comparing two different screening modalities: annual CT scans or chest X-rays in 55,000 smokers and ex-smokers, which they expected to provide more reliable data.[96] Henschke and her co-authors disputed the notion that a randomized controlled trial is necessarily the best approach for evaluating screening.[97] Moreover, they argued that many of the criticisms of their study, such as unnecessary radiation exposure or overdiagnosis applied equally to the NCI trial.

The New York branch of the ELCAP programme, NY-ELCAP received much of its funding out of New York's Tobacco Settlement Fund. ELCAP's critics, not surprisingly, viewed funding this lung cancer screening study as a bad use of public assets.[98] In 2008, even more controversially, another link with tobacco money was revealed in an article in the *New York Times*. The ELCAP study had been financed in part by a Foundation for Lung Cancer: Early Detection, Prevention & Treatment. Henschke was its president and Yankelevitz its secretary-treasurer, and the foundation was reportedly almost entirely underwritten by grants from the parent company of the Ligget group, an American cigarette manufacturer.[99] According to Henschke and Yankelevitz, they had always been open about these grants, which constituted only a small part of the study's overall cost. The outraged responses to the news about the Ligget's grants illustrate my earlier points about stigma. Cancer researchers and journal editors were appalled by Henschke's association with the tobacco industry. The chief medical officer of the American Cancer Society (ACS) is quoted as saying: 'If you're using blood money, you need to tell people you're using blood money'.[100] The ACS would not have supported Henschke's research had it known about the grants from Ligget's. The executive director of the IASLC regretted that research on spiral CT screening was now tainted. The results of the NCI trial comparing screening using conventional chest X-rays and spiral CT were published in December 2011, and indeed, the researchers found that the mortality rate was lower in the group screened by CT.[101] But had there ever been any doubt that CT was a more sensitive technology than chest X-rays? This still did not necessarily mean that a CT screening programme would be cost effective, which was an important consideration.[102]

Whatever the outcome of the controversy over the ELCAP and other screening studies, it is clear that in order to make a screening programme cost effective, minimize the risk of false-positive findings and exclude individuals who would not benefit from interventions, the patients for screening will have to be pre-selected very carefully.[103] The difference between past approaches to screening and the Liverpool Lung Project of the Roy Castle Foundation is that the former dealt with means of detecting tumours that are small and localized enough to allow surgery, while the latter aims at finding means of stopping cells from turning malignant in the first place. Another goal is to identify high risk individuals more reliably.[104] The Project, which Field terms a 'molecular epidemiological study of early lung cancer detection' is under way, but it is too early to say how successful it will be.[105] The Project has two components: a case-control study of 800 newly diagnosed cases, which is meant to enable Field and his colleagues to formulate a molecular genetic and epidemiological model for assessing individual risks of developing lung cancer, and a prospective cohort study with 7,500 high-risk individuals in the Merseyside area over ten years to test, strengthen and confirm the model developed in the case control study.[106] This is a plan for a fairly conventional epidemiological study combined with modern molecular biological methods. But maybe this is exactly what Donnelly wanted to see implemented. What stands out again, however, is the local focus of the study on Liverpool and Merseyside, and the way it is embedded in local networks in which Donnelly and Field had worked for years and which had been strengthened by the early campaigns of the charity.

Conclusion: De-stigmatizing lung cancer

Charities need charisma and rely on the convictions of their supporters and on photogenic (and telegenic) patrons, but they also have to report to the Charity Commission and collaborate with NHS trusts and health bureaucracies.[107] After a period of rapid expansion and faced with recurring fixed costs of its Research Centre and a growing staff, adding to a deficit of several hundred thousand pounds per annum, the Roy Castle Foundation entered a period of crisis in 1999, which lasted into the early years of the new millennium. The Foundation's charismatic chief executive, who had been originally approached to join the Foundation as chief fund raiser and saw fundraising as her main role and strength, resigned after a row with the chairman of trustees and amidst allegations that she had charged the charity for private expenses.[108] Following this crisis, a new chief executive was appointed and new organizational and fundraising models implemented (less spectacle, more sustainability). In 2003, Liverpool University, which previously had been criticized for a lack of co-ordination and investment in its cancer research programmes, purchased the Foundation's headquarters, the Roy Castle Centre, with its brand new laboratory facilities, preserving the Foundation's independence and allowing it to continue its activities as before.[109] There are many aspects of the story of the Roy Castle Foundation that resonate with the history of patient activism around HIV-AIDS or breast cancer, which both helped to reduce the stigma associated with these diseases and lobbied for basic research.[110] Thanks partly to the work of patient activists and cancer charities, such as the Roy Castle Foundation it is ok today to talk about cancer and share personal experiences. Besides research, the Foundation funds and organizes smoking cessation pro-grammes and programmes that help patients cope with the disease and the treatment. In light of the available statistics, some may view the emphasis on survivorship as an expression of the 'tyranny of cheerful-ness' that King and Ehrenreich criticize: the great majority of patients diagnosed with lung cancer will be killed by the disease, as was Roy Castle. Others will welcome some positive spin if it contributes to de-stigmatizing lung cancer.

8
Still Recalcitrant? Some Conclusions

Lung cancer, as I have shown in this book, is more than one disease. Statistically, lung cancer remains the main cause of cancer deaths in Britain, the United States and elsewhere, and while mortality and incidence trends are pointing downward for the industrialized west, this is not the case in developing countries. Lung cancer is the tenth most common cause of death worldwide, and epidemiologists expect it to move up rather than down.[1] Lung cancer is also more than one disease where its biology and natural history are concerned: as cancer specialists know, no tumour is like the other and all patients are different. But as a historian I have been interested in the meanings of the disease. I have shown in this book how multiple identities of lung cancer have emerged over the past two centuries. In the nineteenth century it has emerged as a specific, local disease of the lung (rather than a non-specific fever), and then a disease of cells. In the early twentieth century it turned into a disease treated mostly by surgeons, who could operate on an open thorax only when anaesthetists had developed the technology. Over the course of the twentieth century it acquired the image of a condition where modern treatment modalities such as radiotherapy and chemotherapy commonly failed to save patients' lives, with lung cancer sufferers constituting the largest contingent of cancer patients in hospices. Most prominently, as a result of epidemiological studies and public health policy, this is a disease firmly associated with the habit of smoking cigarettes. I have mostly dealt with Britain and to some degree the United States in this book; the history of this recalcitrant disease in other parts of the world is an important story that will add further layers of meaning but needs to be told by somebody else.

The history of lung cancer, like that of other cancers has been one of increasing differentiation and classification in changing reference frameworks. In Chapter 2 I have shown how the new framework of pathological anatomy turned cancer of the lung into a specific disease: a specific form of consumption, different from tuberculosis. Cellular pathology and the beginnings of modern biomedicine added another level of differentiation: it gave us the notion of transformed, malignant cells, and with the help of microscopes and staining techniques it was now possible to distinguish tumours by assigning them to one of several cell types. These new diagnostic powers, however, made little difference to the treatment of cancer. As I have shown, doctors still employed many remedies that were assumed to work in terms of older paradigms.

The treatment modalities of the twentieth century added another level of difference: cancers which were or were not operable or did or did not respond to radiotherapy or the various cytotoxic drugs. These differences were formalized with the help of standardized classification systems such as those I have discussed in Chapter 6: cell typing to classify the tumours and staging to classify the patients, or rather, those aspects of the patients deemed relevant for the outcome of cancer treatments. This process of increasing differentiation is far from over as new cancer drugs become available that show effects in ever smaller and more specific patient groups. There may well never be a clear-cut cure for lung cancer, a single magic bullet, but there will probably be an increasing number of therapies that make some difference for some subgroups of patients. Given the expense associated with such new, highly specific cancer drugs in recent years, this raises issues of affordability, but also of effective diagnosis and referral. The administrative challenge is greater than one might think: there are still patients who would benefit from a lung resection who fail to be referred to a specialist chest surgeon, as demonstrated by recent lung cancer audits in the UK.[2]

The new drugs are too recent to be covered in this book. For all intents and purposes lung cancer is still a predominantly surgically defined disease, and I have discussed in Chapter 3 how this came to be. With surgery established as the main treatment modality, experiments with other modalities were only possible where surgery was not an option, as became clear during the laborious discussions of the MRC lung cancer working party which I have sketched in Chapter 5. These preceded a set of well-organized but very disappointing trials, adding to a sense of gloom about the treatment of lung cancer, which was taking hold among chest surgeons in the 1960s. This gloom has persisted, as I have argued in my analysis of Ray Donnelly's motives for launching his Lung Cancer Fund in the early 1990s shows. He launched a specific lung cancer charity to bring the progress that he felt had been denied to lung cancer patients because smokers were increasingly seen as undeserving. At a time when 'patients' were becoming the rhetorical focus of public medicine (and not only as consumers), the story of the charity resonated with a broader narrative of regeneration and hope, especially in the Roy Castle Foundation's home town, Liverpool.

The central issue of this book has been recalcitrance. What made lung cancer recalcitrant? If its history made lung cancer into a recalcitrant disease, as I argue in this book, we may want to ask, in turn, what makes some other cancers not recalcitrant. Lung cancer is recalcitrant, for example, because improvements in surgical practice have not made much difference to survival rates for nearly as long as surgery has been established as the main stay of treatment. As I noted in Chapter 5, John Bignall commented in the 1960s, that the course of lung cancer was governed by three main factors: the malignancy of the tumour, its situation in the lung and the age of the patient. All of these, he suggested, were uncontrollable. No surprise, then, that surgeons emphasized the importance of good diagnosis: there was no need to expose patients to surgery who had no hope of benefiting. A similar argument, however, could be made for breast cancer, historically the most visible of all cancers and the one about whose history we know most.[3] Breast cancer is no longer viewed as recalcitrant since the radical mastectomy has become a standard treatment in the early twentieth century. It is, of course, an undisputable statistical fact that outcomes have greatly improved for breast cancer, in ways that lung cancer outcomes have not. But pink ribbons and the gospel of survivorship have rendered somewhat marginal the approximately 15 percent of breast cancer patients who do not survive the disease. Believers in the power of cancer research may want to argue that these represent a remnant from the corners of cancer's history which progress has not reached. The geneticist Maurice Fox, however, suggested in 1979 that there were at least two varieties of breast cancer: one that by histological criteria was malignant but clinically relatively benign, where treatment led to life expectancies similar to women who never encountered the disease, even if this treatment was less radical than customary at the time, and a second variety, which we may want to characterize as recalcitrant, where no treatment made much difference for a patient's likelihood of surviving the disease.[4] While lung cancer came to be defined by its biological recalcitrance, the public image of breast cancer is, as Barbara Ehrenreich put it, increasingly 'sugar coated'.[5] This renders less visible what Robert Aronowitz has observed, that many breast cancer 'survivors' would be more adequately characterized as chronic illness patients, due to adverse treatment effects and the continuing perception that they are at risk.[6]

The recalcitrant image of lung cancer, of course, as I have argued throughout the book, is not only due to biological factors but also the changing image of lung cancer patients and the stigma of a self-inflicted disease associated with a morally somewhat dubious habit. Breast cancer has never carried similar moral stigma, beyond vague feelings of guilt and diffuse concerns that certain behavioural patterns may predestine women to develop cancer. While most smokers have known that smoking was not good for them since the 1950s, even overestimated the likelihood that they would develop lung cancer (see Chapter 7), many defied the warnings of public health campaigners and continued to smoke. But if people continued to smoke against better judgement and conducted a deadly cancer as a consequence, did they not bring it upon themselves? Many in the more affluent and better educated layers of society thought so and did indeed stop smoking. When smoking increasingly acquired associations with membership in lower social strata (and to some degree with delinquency), so did lung cancer by implication.

If lung cancer was recalcitrant, does this mean that it was also neglected? To be sure, lung cancer research is less well funded, but there are other reasons why leukaemia has received more attention from clinical researchers since the 1950s, and why research on and care for childhood cancers has received more charity funding. When the new, highly experimental practices around chemotherapy were developed for the treatment of blood and lymph cancers in the 1950s and 1960s, there was nothing that could be done for leukaemia patients, while there already was a well established standard therapy for lung cancer: surgery. There were also practical advantages that made blood and lymph cancers suitable for these experimental approaches: the effects of the drugs on transformed cells in the blood were infinitely easier to monitor than for solid cancers. Nevertheless, as I have shown in Chapter 5, lung cancer did then receive as much, if not more attention from the MRC than other cancer types. And is it because of neglect that charities founded to help children suffering from leukaemia have received more funding, and much earlier than lung cancer charities? The death of a child (from malignant disease or other causes) strikes more people as cruel and untimely than the cancer death of an elderly man, making it easier to mobilize funding. The particular circumstances of each death are a different matter: most deaths or pathways leading to death are filled with sorrow and confront individuals with situations where they struggle to maintain their dignity.

But there are other, more practical reasons explaining the dominance of some malignancies in basic cancer research. Laboratory-based cancer research since the Second World War has been aimed at illuminating the fundamental molecular mechanisms underlying the cellular transformations that cause many cancers. It has just happened to predominantly rely on tools – tumour cell lines, for example – which have been derived from blood and lymph cancers. Does this mean that this research is leukaemia research? Probably not, unless researchers are explicitly motivated by clinical issues. One might well ask whether molecular biologists who justify their research with the possible implications for the treatment of cancer, are really interested in cancer. This is a question for a different type of study, which does not only concern lung cancer. While in practice often aimed at understanding the molecular mechanisms of life, cancer research has commonly been justified with the vague promise of a magic bullet that enables us to deal with malignancies in the same way as penicillin and other antibiotics have helped us conquer bacterial infections. Great expectations are bound to lead to disappointment and contribute to notions of recalcitrance.

What responses to recalcitrance have I identified in this book? The most common – and default – response was and still is to soldier on, ignore the possibility of failure, keep trying, use the whole arsenal of available treatment modalities: if the default treatment does not work, try something else, which may have worked elsewhere, and never admit that there is nothing else you can do to prolong a patient's life. This approach is fundamental to modern medicine and only rarely criticized. It characterizes treatment policies both on a macro and a micro level: resignation to the inevitable is usually not an option, and where it is, this is often decided quietly, in secret. When we go and see a doctor, we go for a reason: we want to be helped, and we expect that we will benefit from the myriad innovations modern medical science has produced. Not many cancer patients will follow the example of the Catholic priest and critic of many institutions of modernity (including medicine), Ivan Illich, who refused treatment for his parotid cancer, managing the pain with the help of opium and living with the growth for nearly a decade. For most cancers and most patients, there is the expectation that the whole armamentarium of modern, biomedical cancer therapy will be applied, even if the gains in terms of survival time are not particularly impressive. Only when biomedicine runs out of options, do most patients tend to turn to alternative medicine (an issue I have not dealt with in this book) or palliative care. Then the palliative care applied by specialists increasingly turns into one of the standard options, and alternative therapies turn complementary, included in the armamentarium of biomedicine to help patients deal with the experience that their bodies are disintegrating as an effect of surgery, radiotherapy or chemotherapy.

While many other cancers may be recalcitrant, lung cancer is exceptional because of its link with smoking. As I have shown in Chapter 5, the link with smoking provided frustrated trial organizers with a way of rationalizing the failure of their trials: this was a disease that could not be treated; it had to be prevented. So, ironically, knowing the cause of this cancer has not helped sufferers at all. It may have helped decision makers in governments, research institutions and funding agencies to divert funds away from this recalcitrant and apparently self-inflicted disease towards other targets that promised better rewards. But lung cancer does need funding. It needs it in areas that may appear somewhat boring, compared to the excitement associated with cutting edge biomedical research. There may not be a cure in sight (at least not as we imagine it, for the majority of lung cancer patients), but we can invest in care and improving referral pathways, make sure that more patients benefit from best current practice, increase access to long-established treatment modalities or good palliative care, or even grant responsible access to drugs that help patients end their lives as and when they desire. All these will make a difference to many individual lives.

Notes

Chapter 1   Introduction: The History of a Recalcitrant Disease

1   'Henry Marshall Hughes'.
2   Hughes, 'Cases of Malignant Disease of the Lung'.
3   Ibid., 333. Books and articles on physical examination were full of detailed descriptions of specific sounds.
4   Ibid.
5   Ibid., 335.
6   Craig, *The Last Freedom*. More on this story in Timmermann, 'Running Out of Options'.
7   Craig, *The Last Freedom*, 108.
8   On cancer clinical trials, see Keating and Cambrosio, *Cancer on Trial*. For an engaging account of the challenges and successes in the treatment of cancer see Mukherjee, *The Emperor of All Maladies*. On childhood cancer, see Krueger, *Hope and Suffering*; Laszlo, *The Cure of Childhood Leukemia*; Barnes, 'Between Remission and Cure'; Barnes, 'Caring and Curing'.
9   On treatment outcomes, see Macmillan Cancer Support, 'The Cancer Survival Lottery'. On the history of breast cancer: Lerner, *The Breast Cancer Wars*; Aronowitz, *Unnatural History*.
10   Lenfant, 'Shattuck Lecture: Clinical Research to Clinical Practice – Lost in Translation?'. This resonates with the by now classic argument by Thomas McKeown, that medical advances in a narrow sense played only a very minor role in the increase of life expectancy in the developed world over the last two centuries: McKeown, *The Role of Medicine*.
11   Taylor, 'Clinical Lecture', 875, 877.
12   Cf. Pickstone, 'Contested Cumulations'.
13   Thatcher et al., 'Chemotherapy in Non-small Cell Lung Cancer', 85.
14   This is the figure for 2008, the most recent year for which worldwide data is available: World Health Organization, 'Fact Sheet No 297: Cancer'; for the UK and the US, see Cancer Research UK, 'Lung Cancer and Smoking'; Centers for Disease Control and Prevention, 'CDC – Lung Cancer Statistics'.
15   Cancer Research UK, 'Pancreatic Cancer Statistics'; Cancer Research UK, 'Liver Cancer Statistics – UK'. On pancreatic cancer, see also the patient memoir by Pausch, *The Last Lecture*.
16   Taylor, *Smoke Ring*. See also, for example, Brandt, *The Cigarette Century*; Berridge, *Marketing Health*; Berlivet, 'Association or Causation?'; Lock et al., *Ashes to Ashes*; Burnham, 'American Physicians and Tobacco Use'.
17   See also Brandt and Rozin, *Morality and Health*.
18   Holmes Sellors, 'The Management of Cancer of the Lung', 19.
19   Cancer Research UK, 'Lung Cancer and Smoking'; Centers for Disease Control and Prevention, 'CDC – Lung Cancer Statistics'.

20　Kmietowicz, 'Research Spending on Cancer Doesn't Match Their Death Rates'. See also Chapter 7.
21　Timmermann, 'As Depressing as It Was Predictable'.
22　The idea that it was important to diagnose early and not to delay treatment had long been associated with breast cancer and increasingly with other female cancers: Aronowitz, 'Do Not Delay'; Aronowitz, *Unnatural History*; Lerner, *The Breast Cancer Wars*; Löwy, *Preventive Strikes*.
23　Belcher, 'Indications for Surgery and Choice of Operation', 100.
24　Thatcher et al., 'Chemotherapy in Non-small Cell Lung Cancer'. See also the discussion documented on pp. 94–95 – the paper was originally presented during a symposium.
25　Cf. Aronowitz, 'Do Not Delay'.
26　Nathoo, *Hearts Exposed*; Lederer, *Flesh and Blood*.
27　Cf. Donnelly, *Cinderella Cancer*.
28　See, for example, Keating, *Smoking Kills*.
29　Mukherjee, *The Emperor of All Maladies*.
30　See Keating and Cambrosio, *Cancer on Trial*; Barnes, 'Between Remission and Cure'.
31　Wishart, *One in Three*.
32　Szabo, *Incurable and Intolerable*.
33　Examples from sociology include Charmaz, *Good Days, Bad Days*; Frank, *At the Will of the Body*; Frank, *The Wounded Storyteller*; Kleinman, *The Illness Narratives*.
34　See, for example, Berridge, *Marketing Health*; Brandt, *The Cigarette Century*; Berridge, *Making Health Policy*; Keating, *Smoking Kills*; Abbey Smith, 'Development of Lung Surgery'; Mueller, *Evarts A. Graham*.
35　Cf. Berridge, *Making Health Policy*; Berridge, *Marketing Health*.
36　Valier and Timmermann, 'Clinical Trials'.

Chapter 2　Lung Cancer and Consumption in the Nineteenth Century: Bodies, Tissues, Cells and the Making of a Rare Disease

1　Adler, *Primary Malignant Growths*. He had published on the subject earlier: Adler, 'The Diagnosis of Malignant Tumors of the Lung'. For contemporary accounts of lung cancer, see also the useful article by Rosenblatt, 'Lung Cancer in the 19th Century'. Less useful: Onuigbo, 'Lung Cancer in the Nineteenth Century'.
2　Adler, *Primary Malignant Growths*, 3.
3　Pässler, 'Ueber das primäre Carcinom der Lunge'.
4　Maulitz, *Morbid Appearances*.
5　On the history of this liaison, see Lawrence, *Medical Theory, Surgical Practice*.
6　See Maulitz, *Morbid Appearances*.
7　For an excellent study on Laennec, see Duffin, *To See with a Better Eye*. On medicine in revolutionary and post-revolutionary Paris, see Ackerknecht, *Medicine at the Paris Hospital*; Foucault, *The Birth of the Clinic*. On the reception of the new pathological anatomy, see Maulitz, *Morbid Appearances*.
8　For other examples of disease histories, see Rosenberg and Golden, *Framing Disease*.

9 On the history of auscultation and the impact of the new technique on the medical encounter, see also Lachmund, *Der abgehorchte Körper*.

10 René T. H. Laennec, 'Encéphaloides', in *Dictionaire des Sciences Médicales, par une société de médecins et de chirurgiens* (Paris: C. L. F. Panckoucke, 1815), 165–178.

11 See Laennec, *A Treatise on the Diseases of the Chest*, 137–145.

12 Duffin, *To See with a Better Eye*, 61.

13 Ackerknecht, 'Historical Notes on Cancer'.

14 Laennec, *A Treatise on the Diseases of the Chest*.

15 On Forbes, see Agnew, *The Life of Sir John Forbes (1787–1861)*.

16 Cf. Duffin, *To See with a Better Eye*, 211–213; Maulitz, *Morbid Appearances*, 168–170.

17 Maulitz, *Morbid Appearances*, 134.

18 See also Cunningham, *The History of British Pathology*.

19 'Middlesex Hospital: Case of a Wound Penetrating the Cavity of the Thorax', 95.

20 'The Late Dr Warren', 550.

21 Forbes, *Original Cases*.

22 See Cunningham, *The History of British Pathology*, 31–99.

23 Stokes, *Treatise*, chapter on 'Cancer of the Lung', 370–388.

24 Ibid., 370.

25 Ibid., 371.

26 Ibid., 373.

27 Ibid., 374.

28 Ibid., 375.

29 Ibid., 376.

30 Ibid.

31 Ibid., 376–377.

32 Ibid., 380.

33 Carswell, a Scotsman, had studied in France (with Laennec among others) and had been appointed in 1828 as first Professor of Pathological Anatomy in England at the medical school of the University of London (later University College Medical School). See Maulitz, *Morbid Appearances*, 215–223.

34 Taylor, 'Clinical Lecture'.

35 Ibid., 873.

36 Ibid., 877.

37 Ibid.

38 Ibid., 874.

39 Ibid., 875.

40 Ibid.

41 Ibid., 876.

42 Walshe, *The Nature and Treatment of Cancer*.

43 Taylor, 'Clinical Lecture', 876.

44 A detailed discussion provides, for example, Rather, *The Genesis of Cancer*.

45 See Otis, *Müller's Lab*.

46 The achromatic objective was introduced in 1826 and immediately improved by Lister and many others, turning compound microscopes into reliable and trusted instruments. See Wilson, 'Virchow's Contribution to the Cell Theory'. For the implications of the new microscopes for pathology, see also Foster, *A Short History of Clinical Pathology*.

47   Ackerknecht, 'Historical Notes on Cancer', 118.
48   Virchow, *Die Cellularpathologie*; Virchow, *Die krankhaften Geschwülste*.
49   Ackerknecht, 'Historical Notes on Cancer', 118. See also Plaut, 'Rudolf Virchow and Today's Physicians and Scientists'.
50   Gruhn, 'A History of the Histopathology of Lung Cancer', 28.
51   Adler, *Primary Malignant Growths*.
52   The history of nineteenth century ideas about cancer causation is too complex to be discussed here in great detail. For a comprehensive history of these ideas, see Wolff, *Die Lehre von der Krebskrankheit*.
53   See also Jacyna, 'The Laboratory and the Clinic'.
54   Adler, *Primary Malignant Growths*, 14.
55   Ward, 'Seamen's Hospital, "Dreadnought": Medullary Cancer of Mediastinal Glands and Left Lung'.
56   Ibid., 238.
57   Charteris and Williams, 'Glasgow Royal Infirmary: Cancer of Lung and Pleuro-pneumonia'.
58   Pitt, 'Malignant Disease of Bronchial Glands'.
59   Gruhn, 'A History of the Histopathology of Lung Cancer', 33.
60   Mackenzie, *A Practical Treatise on the Sputum*. See also Walloch, 'Pulmonary Cytopathology in Historical Perspective', 176.
61   Curran, 'A Puzzling Case of Cancer of the Lung'.
62   Ibid., 259.
63   Ibid.
64   Ibid.
65   Ibid.
66   Smithers, 'Facts and Fancies About Cancer of the Lung'.
67   See, for example, Stolley and Lasky, *Investigating Disease Patterns*, chap. 2; Lilienfeld and Stolley, *Foundations of Epidemiology*, chap. 2.
68   Eyler, *Victorian Social Medicine*.
69   Peitzman, *Dropsy, Dialysis, Transplant*.
70   King and Newsholme, 'On the Alleged Increase in Cancer'.
71   On Newsholme, see Eyler, *Sir Arthur Newsholme and State Medicine*.
72   King and Newsholme, 'On the Alleged Increase in Cancer', 209.
73   Ibid.
74   Newsholme, 'The Statistics of Cancer'.
75   Eyler, *Sir Arthur Newsholme and State Medicine*, 31.
76   Williams, *The Natural History of Cancer*, 52.
77   Ibid., 62.
78   Table XIII, King and Newsholme, 'On the Alleged Increase in Cancer', 239.
79   Adler, *Primary Malignant Growths*, 3.
80   Ibid.
81   Ibid.
82   Ibid., 10–11.
83   Ibid., 11–12.
84   Lerner, *The Breast Cancer Wars*, 42.

Chapter 3 Lungs in the Operating Theatre, circa 1900 to 1950

1 Adler, *Primary Malignant Growths*, 11–12.
2 Cf. Lerner, *The Breast Cancer Wars*; Aronowitz, *Unnatural History*.
3 Lawrence, 'Democratic, Divine and Heroic', 32–33. For an enthusiastic account of early twentieth century developments in surgery, see Slaughter, *The New Science of Surgery*. There are some very good insider histories of thoracic surgery available. See, for example, Hurt, *The History of Cardiothoracic Surgery*; Naef, *The Story of Thoracic Surgery*; Naef, 'The Mid-century Revolution in Thoracic and Cardiovascular Surgery: Parts 1–6'; Abbey Smith, 'Development of Lung Surgery'; Meade, *A History of Thoracic Surgery*. Especially on lung cancer, see: Mountain, 'The Evolution of the Surgical Treatment of Lung Cancer'; Rubin, 'Lung Cancer: Past, Present, and Future'; Brewer, 'Historical Notes on Lung Cancer before and after Graham's Successful Pneumonectomy in 1933'.
4 Bronchiectasis and empyema are diseases classified according to clinical observations and autopsy findings. Empyema, in essence, is the accumulation of liquid in the space between lung and ribcage, usually caused by an infection, for example streptococcal pneumonia. Bronchiectasis is the term given to a set of symptoms first described by Laennec, which include the collapse, widening and fusion of alveolar sacs, the finest branches of the airways. These abnormal cavities become infected and fill with sputum, often smelling foul, which patients struggle to cough up. Bronchiectasis can be a consequence of an inherited defect or caused by infections.
5 Lawrence, 'Democratic, Divine and Heroic', 23–28.
6 Meade, *A History of Thoracic Surgery*, 28–97.
7 Lawrence, 'Democratic, Divine and Heroic', 28–32.
8 Evarts Graham uses these terms, for example in Graham, 'Changing Concepts in Surgery'.
9 Cf. Abbey Smith, 'Development of Lung Surgery'.
10 'Memorandum by the Society of Thoracic Surgeons of Great Britain and Ireland'.
11 Paneth, 'The Brompton Hospital and Cardiothoracic Surgery'.
12 Bryder, *Below the Magic Mountain*.
13 Mason, 'The Surgical Treatment of Pulmonary Tuberculosis'.
14 Davidson and Rouvray, *The Brompton Hospital*, 107.
15 Brompton Hospital, 'Medical Report for 1914'. See also Brock, 'Thoracic Surgery'.
16 Mushin and Rendell-Baker, *The Principles of Thoracic Anaesthesia*.
17 Image source: Sauerbruch, 'Zur Pathologie des offenen Pneumothorax'.
18 See Brodsky, 'The Evolution of Thoracic Anesthesia'; Naef, *The Story of Thoracic Surgery*, esp. 4–9.
19 Davies, 'The Mechanical Control of Pneumothorax'.
20 Image source: Davies, 'The Mechanical Control of Pneumothorax'.
21 Davies, 'Recent Advances in the Surgery of the Lung and Pleura'. See also Hurt, *The History of Cardiothoracic Surgery*, 271–272.

22   Heroic firsts are a problem in the history of surgery. In order for an operation to be counted as a success, the patient did not necessarily have to
survive it for very long. In the literature on the subject written by in-
siders, one finds several first lobectomies. Scannell, for example, credits
the American surgeon Samuel Robinson, the fourth president of the
AATS, with performing a lobectomy for bronchiectasis in 1909.
Admittedly, this was a 'multistaged affair', but the patient recovered.
Scannell, 'Historical Perspectives of the American Association for
Thoracic Surgery: Samuel Robinson (1875–1947)'.

23   Webb cites an occasion when the President of the London Medical
Society stopped Morriston Davies with the words, 'This has got nothing
to do with the treatment of tubercle'. Webb, *Hugh Morriston Davies*, 17.

24   Fellow chest surgeon H. P. Nelson, consultant to the Brompton, the
Papworth Hospital in Cambridge and the London Hospital, died in 1936
from a streptococcal septicaemia that he had contracted while dressing
an empyema wound.

25   We can only speculate what this meant for patients.

26   Morriston Davies's life and career is well documented, thanks to the
efforts of Kathleen Webb, an amateur historian whose father used to run
the garage in Ruthin and look after Morriston Davies's car: Webb, *Hugh
Morriston Davies*. See also 'Dr Hugh Morriston Davies Conversation'. This
document is the transcript of a conversation between Morriston Davies
and Sam Jones and Dyfnallt Morgan of the BBC, in preparation for a
broadcast portrait of Morriston Davies in autumn 1960, a recording of
which I have not been able to locate. Included in the file is a detailed CV
of Morriston Davies. Unfortunately the conversation is rather brief on
Morriston Davies's later career in Liverpool. The *Lancet* published a long
obituary: 'Obituary: Hugh Morriston Davies'.

27   See Brodsky, 'The Evolution of Thoracic Anesthesia'; Naef, *The Story of
Thoracic Surgery*, esp. 4–9.

28   Magill et al., 'Lest We Forget', 477; see also Magill, 'Endotracheal
Anesthesia'; 'An Appraisal of Progress in Anaesthetics'.

29   Magill et al., 'Lest We Forget', p. 477.

30   Magill, 'Anaesthesia in Thoracic Surgery'.

31   Nosworthy, 'Anaesthesia in Chest Surgery'.

32   Davies and Coope, *War Injuries of the Chest*.

33   Olch, 'Evarts A. Graham in World War I'. See also Mueller, *Evarts
A. Graham*.

34   Robinson, 'The Present and Future in Thoracic Surgery', 250.

35   Ibid.

36   Ibid.

37   Lilienthal, 'Resection of the Lung', 257. Quoted, for example, in Hurt,
The History of Cardiothoracic Surgery, 225.

38   The technique was developed by Gustav Killian in Freiburg, Germany, in
the late nineteenth century. Cf. Hurt, *The History of Cardiothoracic
Surgery*, 49–53.

39   Lilienthal, 'Resection of the Lung'.

40   Ibid., 260.

41   Davies, 'Recent Advances in the Surgery of the Lung and Pleura', 254.

42 See, for example, Hurt, *The History of Cardiothoracic Surgery*, 271–272; Meade, *A History of Thoracic Surgery*, 60–62.

43 Davies, 'Recent Advances in the Surgery of the Lung and Pleura', 256.

44 Mason, 'Extirpation of the Lung', 1048.

45 Cf. Mackenzie, *Cancer*, esp. 16–17.

46 Meade, *A History of Thoracic Surgery*, 84. The story is told in detail in Mueller, *Evarts A. Graham*, 117–140.

47 Hurt, *The History of Cardiothoracic Surgery*, 273.

48 Ibid.

49 Logan, 'The Beginnings of Thoracic Surgery'. See also Mason, 'Extirpation of the Lung'.

50 Logan's account of the operation inspired Hurt to marvel about the remarkable healing power of the human body: Hurt, *The History of Cardiothoracic Surgery*, p. 283.

51 Mueller, *Evarts A. Graham*.

52 Ibid.; Graham and Singer, 'Successful Removal of an Entire Lung for Carcinoma of the Bronchus'.

53 Mueller, *Evarts A. Graham*, 130.

54 Hurt, *The History of Cardiothoracic Surgery*, 273.

55 Crafoord, *On the Technique of Pneumonectomy in Man*.

56 On Roberts, see 'Obituary: J. E. H. Roberts'; 'Obituary: James Ernest Helme Roberts'. On Tudor Edwards: 'Obituary: Arthur Tudor Edwards'; 'Tudor Edwards Memorial', 7 June 1947; 'Tudor Edwards Memorial', 13 November 1948; 'In Memoriam: Arthur Tudor Edwards (1890–1946)'. The relationship between the two is said to have been problematic. Cf. Hurt, *The History of Cardiothoracic Surgery*.

57 Hurt, *The History of Cardiothoracic Surgery*, 467.

58 See ibid., 267–295, for a detailed history of the operations.

59 'Obituary: Arthur Tudor Edwards', 365.

60 Smithers, 'Facts and Fancies About Cancer of the Lung', 1235–1236.

61 Cf. Jacyna, 'The Laboratory and the Clinic'.

62 See Brompton Hospital, 'Medical Report for 1906', BH/A/13, Royal London Hospital Archives.

63 Davidson and Rouvray, *The Brompton Hospital*, 118.

64 Ibid., 119.

65 Brompton Hospital, 'Medical Report for 1920', BH/A/13, Royal London Hospital Archives.

66 Brompton Hospital, 'Medical Report for 1921', BH/A/13, Royal London Hospital Archives.

67 Brompton Hospital, 'Medical Report for 1922', BH/A/13, Royal London Hospital Archives.

68 Brock, 'Thoracic Surgery'. Brock's numbers are slightly different from those in the Medical Reports, and the series is not complete.

69 Ibid.

70 Abbey Smith, 'Development of Lung Surgery'.

71 Ibid. A bronchopleural fistula is a hole, after failure of the bronchial stump to heal (often because of an infection), allowing air to flow between bronchus and pleural space.

72 Cooter, 'Keywords in the History of Medicine: Teamwork'.

73   'Obituary: Arthur Tudor Edwards', 365.
74   Ibid.
75   Bryce, 'Letter to Hugh Morriston Davies'; Bryce, 'Letter to J. E. H. Roberts', 2 November 1931, Bryce Papers, MS0005, RCS Archives.
76   Bryce, 'Letter to J. E. H. Roberts', 9 May 1933, Bryce Papers, MS0005, RCS Archives.
77   When the British Association were discussing the subscription to the AATS's *Journal of Thoracic Surgery*, Henry Nelson wrote to Bryce that he felt that 'the journal is much too full of experimental animal surgery and not enough clinical'. Nelson, 'Letter to Alexander Graham Bryce', Bryce Papers, MS0005, RCS Archives.
78   Scannell, 'Historical Perspectives of the American Association for Thoracic Surgery: Samuel J. Meltzer (1851–1920)'.
79   Ibid., 905.
80   Allison, 'Letter to Arthur Tudor Edwards'; 'Circular', Bryce Papers, MS0005, RCS Archives.
81   Society of Thoracic Surgeons of Great Britain and Ireland, 'Memorandum on the Provision of a National Thoracic Surgery Service', November 1943, Bryce Papers, MS0005, RCS Archives.
82   Society of Thoracic Surgeons of Great Britain and Ireland, 'Memorandum on the Provision of a National Thoracic Surgery Service', March 1948, copy in the author's possession.
83   'A National Thoracic Surgical Service (editorial)'.
84   Ibid., 633.
85   Society of Thoracic Surgeons of Great Britain and Ireland, 'Memorandum on the Provision of a National Thoracic Surgery Service', March 1948, 11.
86   See also 'Tuberculosis and the National Health Service: Report of a B.M.A. Group Committee'.
87   'Memorandum for the Ministry of Health by the Society of Thoracic Surgeons of Great Britain and Ireland', BD 18/901, UK NA.
88   Bryce, 'Letter to J. E. H. Roberts', 21 October 1941, Bryce Papers, MS0005, RCS Archives. On hospitals in Manchester, see Pickstone, *Medicine and Industrial Society*. On the Bagueley, later Wythenshawe Hospital, see also Davies, *Baguley and Wythenshawe Hospitals*.
89   Davies, 'Memories Provided by John Dark'. See also the figures quoted earlier.
90   Nicholson et al., 'Review of 910 Cases of Bronchial Carcinoma'.
91   Interview with Mr John Dark, thoracic surgeon, Wythenshawe Hospital, Manchester.
92   Ibid.
93   Quoted after Davies, *Baguley and Wythenshawe Hospitals*, 93.
94   Howell, 'Soldier's Heart'.
95   Smithers, 'Clinical Cancer Research'.
96   Smithers, *Not a Moment to Lose*, 25. The lack of interest from diagnostic radiologists was fairly common: Cf. Pickstone, 'Contested Cumulations'.
97   Smithers, *Not a Moment to Lose*, 32. Surgeons played a central role in the reform of services in the US; but little is known for Britain. The historiography for the British case is biased towards clinical medicine; good social histories of recent surgery are rare. Cf. Pickstone, 'Contested Cumulations'.

98   Smithers, *Not a Moment to Lose*, 31. Again, this was fairly typical: Cf. Pickstone, 'Contested Cumulations'.
99   Murphy, 'A History of Radiotherapy to 1950'.
100   Smithers, *Not a Moment to Lose*, 25.
101   Wood was a distinguished pioneer of radiology and one of very few women doctors in similar positions in her generation. Halnan, 'Obituary: Dr Constance A. P. Wood'.
102   Smithers, *Not a Moment to Lose*, 26.
103   Ibid., 34.
104   Cf. Pickstone, 'Contested Cumulations'.
105   Smithers, *The X-ray Treatment of Accessible Cancer*. Smithers, Branson, and Hartley, *The Royal Cancer Hospital Mechanically Sorted Punched Card Index System*.
106   The one rejected proposal was to abolish the surgical, one-man cancer clinics and replace them with specialist clinics. It was rejected because it restricted the freedom of consultants to treat any patient referred to them in the way they deemed suitable.
107   Smithers, *Not a Moment to Lose*, 36. On Cade, see Cox, 'In Memoriam: Sir Stanford Cade, KBE CB'.
108   Henk, 'Obituary: Professor Sir David Smithers'; 'Sir David Smithers: Obituary', 19.
109   'Folder: Film – Thoracic Surgery', BW 4/35, UK NA. The folder contains draft scripts and correspondence. A copy of the film can be viewed at the British Film Institute.
110   Some documents also refer to 'Surgery in Chest Diseases' [plural], but this is the same film.
111   Bentley and Leitner, 'Mass Radiography. With Special Reference to Screen Photography and Pulmonary Tuberculosis'.
112   Bundy, 'Memo to Kearney', BW 4/35, UK NA.
113   Elton, 'Letter to A. F. Primrose (British Council)', BW 4/35, UK NA.
114   Primrose, 'Letter to Arthur Elton (Ministry of Information)', BW 4/35, UK NA.
115   'Chest Surgery Film'.
116   Watson, 'Thoracic Surgical Service 1940–1941', 56.
117   Timmermann, '"Just Give Me the Best Quality of Life Questionnaire"'.
118   Keating and Cambrosio, *Cancer on Trial*.

Chapter 4   Science, Medicine and Politics: Lung Cancer and Smoking, circa 1945 to 1965

1   Hilton, *Smoking in British Popular Culture*, 180.
2   Rothstein, *Public Health and the Risk Factor*; Hill et al., '"The Great Debate"'; Talley et al., 'Lung Cancer, Chronic Disease Epidemiology, and Medicine'; Berlivet, 'Association or Causation?'; Lock et al., *Ashes to Ashes*; Doll, 'Uncovering the Effects of Smoking'; Clemmesen, 'Lung Cancer from Smoking'.
3   Doll and Hill, 'The Mortality of Doctors in Relation to Their Smoking Habits'.

4   For Britain, see Berridge, *Marketing Health*; Taylor, *Smoke Ring*. For the US: Brandt, *The Cigarette Century*.

5   Cole, 'The Economic Effects', 60.

6   Ibid.

7   For Britain, see Austoker, *A History of the Imperial Cancer Research Fund*, especially pp. 162–204; for overviews on cancer in the twentieth century, see Gaudillière, 'Cancer'; Cantor, *Cancer in the Twentieth Century*; Pinell, 'Cancer'; Löwy, 'Cancer'; Cantor, 'Cancer'.

8   Duguid, 'The Incidence of Intrathoracic Tumours in Manchester'; Bonser, 'The Incidence of Tumours of the Respiratory Tract in Leeds'.

9   Duguid, 'The Incidence of Intrathoracic Tumours in Manchester', 111.

10 'Cancer of the Lung (editorial)'.

11 Ibid.

12 Sitsen, 'Über die Häufigkeit des Lungenkrebses'; Sitsen, 'Wird der Lungenkrebs häufiger?'.

13 For the history of the institute, see Brunning and Dukes, 'The Origin and Early History of the Institute of Cancer Research of the Royal Cancer Hospital'; see also Wiltshaw, *A History of the Royal Marsden Hospital*.

14 Kennaway and Kennaway, 'A Study of the Incidence of Cancer of the Lung and Larynx', 266.

15 Ibid., 255. On Ernest Kennaway, see Cook, 'Ernest Laurence Kennaway. 1881–1958'.

16 For figures, see Charlton and Murphy, *The Health of Adult Britain 1841–1994*, 45–46.

17 Joint National Cancer Survey Committee of the Marie Curie Memorial and the Queen's Institute of District Nursing, *Report on a National Survey Concerning Patients with Cancer Nursed at Home*, 22.

18 Image source: Royal College of Physicians, *Smoking and Health*, 15.

19 Image source: Smithers, 'Facts and Fancies About Cancer of the Lung'.

20 Winter, 'Early Symptomatology of Carcinoma of the Lung'. See also Smithers, 'Facts and Fancies About Cancer of the Lung'. Even X-ray images usually did not show tumours in the lung conclusively until they had reached a considerable size: Simon, 'Radiographic Aspects of Carcinoma of the Lung'.

21 Fry, 'Chronic Bronchitis in General Practice'.

22 Daley, 'The Health of the Nation', 1284.

23 Dingle, 'Studies of Respiratory and Other Illnesses in Cleveland (Ohio) Families'.

24 Fletcher et al., 'The Significance of Respiratory Symptoms and the Diagnosis of Chronic Bronchitis in a Working Population'.

25 Fry, 'Chronic Bronchitis in General Practice'.

26 'Another Winter of Bronchitis (editorial)'.

27 Higgins et al., 'Respiratory Symptoms and Pulmonary Disability in an Industrial Town'.

28 Fry, 'Chronic Bronchitis in General Practice'. On the Great Smog: Berridge and Taylor, *The Big Smoke*.

29 Breslow and Goldsmith, 'Health Effects of Air Pollution'.

30 Kotin and Falk, 'Air Pollution and Its Effect on Health'; Barach, 'Air Pollution and Health'.

31 Fletcher, 'Chronic Disabling Respiratory Disease'.
32 Higgins, 'Respiratory Symptoms, Bronchitis, and Ventilatory Capacity in Random Sample of an Agricultural Population'.
33 Fry, 'Chronic Bronchitis in General Practice'; Higgins, 'Tobacco Smoking, Respiratory Symptoms, and Ventilatory Capacity: Studies in Random Samples of the Population'.
34 Hill, 'Obituary: Dr Percy Stocks, 1889–1974'; Reid et al., 'Percy Stocks: An Appreciation'; Doll, 'Stocks, Percy (1889–1974)'.
35 Figures from Kennaway and Kennaway, 'A Further Study of the Incidence of Cancer of the Lung and Larynx'.
36 Smithers, 'Facts and Fancies About Cancer of the Lung'.
37 Kennaway and Kennaway, 'A Further Study of the Incidence of Cancer of the Lung and Larynx'.
38 Farrow, 'Letter to J. E. Pater (Ministry of Health)', FD 1/1989, UK NA. See also Cuthbertson, 'Historical Notes on the Origin of the Association Between Lung Cancer and Smoking'. It is not entirely clear from the correspondence if Stocks and others meant that X-ray examinations caused or just revealed more lung cancer. Probably there was a bit of slippage and both possibilities seemed worth considering.
39 Mellanby, 'Memorandum', FD 1/1989, UK NA.
40 On Bradford Hill, see Doll, 'Austin Bradford Hill, 8 July 1897–18 April 1991'.
41 Hill, 'Letter to Frank H. K. Green', FD 1/1989, UK NA.
42 Stocks, 'Medical Research Council, Cancer of the Lung (Memorandum Prepared by Dr. Percy Stocks)', 10, FD 1/1989, UK NA.
43 'Medical Research Council, Conference on Cancer of the Lung Held on 6 February 1947, Minutes', FD 1/1989, UK NA. The conference participants were not aware either of the Schairer and Schöniger case control study or the work that Wynder and Graham were doing in the US. On Schairer and Schöniger, see Proctor, 'Commentary: Schairer and Schöniger's Forgotten Tobacco Epidemiology'.
44 Hill et al., 'Proposed Investigation of Cancer of the Lung', FD 1/1989, UK NA. For Doll's version of the story, see Doll, 'The First Reports on Smoking and Lung Cancer'; Doll, 'Uncovering the Effects of Smoking'.
45 Mellanby, 'Letter to Austin Bradford Hill'; Hill, 'Letter to Edward Mellanby', FD 1/1993, UK NA. On the MRC Social Medicine Research Unit, see Murphy, 'The Early Days of the MRC Social Medicine Research Unit'.
46 'Medical Research Council, Conference on Cancer of the Lung Held on 29 September 1947, Minutes', FD 1/1989, UK NA.
47 'Folder: Cancer of the Lung: Res. On, E.L. Kennaway', passim, FD 1/1990, UK NA.
48 'Medical Research Council, Conference on Cancer of the Lung Held on 6 February 1947, Minutes', FD 1/1989, UK NA.
49 Doll, 'Cancer of the Lung Investigation: Interim Report', FD 1/1989, UK NA.
50 Wynder and Graham, 'Tobacco Smoking as Possible Etiologic Factor'.
51 Doll and Hill, 'Smoking and Carcinoma of the Lung'.
52 Wynder and Graham, 'Tobacco Smoking as Possible Etiologic Factor'.

53 Brandt, *The Cigarette Century*, 127–128. See also Ochsner, *Smoking and Cancer: A Doctor's Report*.

54 Mueller, *Evarts A. Graham*, 355–357.

55 Ibid., 357–361. Mueller interviewed Wynder for his reconstruction of the events. See also Wynder, 'Tobacco as a Cause of Lung Cancer'. Or the numerous obituaries, for example Doll, 'In Memoriam: Ernst Wynder, 1923–1999'; Hoffmann and Hoffmann, 'Obituary: Ernst L. Wynder MD Dr Sc Hc (mult) Dr Med Hc, 1922–1999'.

56 See Proctor, *The Nazi War on Cancer*. See also the articles in a special issue of the *International Journal of Epidemiology* in 2001, e.g. Proctor, 'Commentary: Schairer and Schöniger's Forgotten Tobacco Epidemiology'; Doll, 'Commentary: Lung Cancer and Tobacco Consumption'; Zimmermann et al., 'Commentary: Pioneering Research into Smoking and Health in Nazi Germany'.

57 Lickint, *Zigarette und Lungenkrebs*; Lickint, *Lungenkrebs der Raucher*.

58 Lickint, *Tabakgenuß und Gesundheit*.

59 Schairer and Schöniger, 'Lung Cancer and Tobacco Consumption'. The paper was originally published in German in *Zeitung für Krebsforschung*, 34 (1943), 261–269.

60 Proctor, 'Commentary: Schairer and Schöniger's Forgotten Tobacco Epidemiology', 32.

61 Ibid.; Smith et al., 'Smoking and Death [letter]'.

62 See, for example, Garfield, 'The Man Who Saved a Million Lives'; Saxon, 'Ernst Wynder, 77, a Cancer Researcher, Dies'.

63 Berridge, *Marketing Health*; Brandt, *The Cigarette Century*; Parascandola, 'Cigarettes and the US Public Health Service in the 1950s'.

64 Berlivet, 'Association or Causation?'; Parascandola et al., 'Two Surgeon General's Reports on Smoking and Cancer'.

65 Doll and Hill, 'The Mortality of Doctors in Relation to Their Smoking Habits'; Doll et al., 'Mortality from Cancer in Relation to Smoking'.

66 Hammond and Horn, 'The Relationship Between Human Smoking Habits and Death Rates'; Garfinkel, 'Classics in Oncology: E. Cuyler Hammond'.

67 Bloor, *Knowledge and Social Imagery*. More accessible: Collins and Pinch, *The Golem*.

68 Brandt, *The Cigarette Century*; Taylor, *Smoke Ring*; Kluger, *Ashes to Ashes*.

69 'Cigarettes and Cancer (editorial)'.

70 'Experimental Links Between Tobacco and Lung Cancer (editorial)'.

71 Doll and Hill, 'The Mortality of Doctors in Relation to Their Smoking Habits'.

72 On cultural and policy implications of smoking in the UK and the US, see Berridge, 'Science and Policy'; Brandt, *The Cigarette Century*; Burnham, *Bad Habits*.

73 Drayson, 'Non-smoking Carriages (Letter to the Editor)', 5.

74 Berridge and Loughlin, 'Smoking and the New Health Education'.

75 On Joules's campaign, see Palladino, 'Discourses of Smoking, Health, and the Just Society'. See also Hilton, *Smoking in British Popular Culture*; Booth, 'Smoking and the Royal College of Physicians'.

76 Joules, 'Liability to Lung Cancer (Letter to the Editor)', 9.

77 Joules, 'Symposium on Cancer of the Lung and Tobacco Consumption', 18–19.
78 Todd, 'Typescript: Smoking and Lung Cancer: A Statistical Survey', FD 1/2009, UK NA.
79 'Smoking and Cancer', 20 February 1954; 'Lung Cancer and Smoking (editorial)'.
80 'Smoking and Health: 4 New Moves Are Forecast. All Out Research', 1.
81 'Smoking and Cancer', 13 February 1954, 7; 'Link Between Smoking and Cancer. Government Acts: Tobacco Firms' Research Offer', 1; 'Smoking and Health: 4 New Moves Are Forecast. All Out Research'.
82 'Smoking and Cancer', 20 February 1954, 523.
83 Taylor, *Smoke Ring*.
84 Quoted after Ibid., 81.
85 Reynolds, 'To Mr I. McLeod, Minister of Health', MH 55/1012, UK NA. Other letters can be found in the same file.
86 Taylor, *Smoke Ring*, 81.
87 'School 1963 – One in Five Are Smokers', 3.
88 'Cigarettes in School Tuck Boxes', 8.
89 Medical Research Council, 'Medical Research Council's Statement on Tobacco Smoking and Cancer of the Lung'; Medical Research Council, 'Tobacco Smoking and Cancer of the Lung: Statement by the Medical Research Council'.
90 'Lung Cancer Increase "Due to Smoking": Medical Findings Accepted by Government. Local Authorities Asked to Make Facts Known', 10.
91 Royal College of Physicians, *Smoking and Health*.
92 Bedford, 'The Case Against Cigarettes', 16.
93 Ibid.
94 Booth, 'Smoking and the Royal College of Physicians'; Booth, 'Smoking and the Gold-headed Cane: The Royal College of Physicians Enters the Modern World'.
95 'No Smoking?', 878.
96 Quoted after Bedford, 'The Case Against Cigarettes', 16.
97 Advisory Committee to the Surgeon General of the Public Health Service, *Smoking and Health*. British smokers were unimpressed: 'Smokers Carry on as Usual', 2. I do not have the space to introduce the US debate here in detail. See Brandt, *The Cigarette Century*; Kluger, *Ashes to Ashes*; Harkness, 'The U.S. Public Health Service and Smoking in the 1950s: The Tale of Two More Statements'; Talley et al., 'Lung Cancer, Chronic Disease Epidemiology, and Medicine'; Burnham, 'American Physicians and Tobacco Use'.
98 'No Smoke Without Harm'.
99 Doll, 'Uncovering the Effects of Smoking', 103.
100 'Smoke Abatement', 15.
101 Ibid., 16.
102 Cole, 'The Economic Effects', 47.
103 '"Ban the Smokers" Bid in Four Cities', 1.
104 'Smoke Abatement', 15.
105 'Cigarettes: No Drastic Steps Likely', 2.
106 'Cigarette Slot Machine Ban', 1.

107   Taylor, *Smoke Ring*, 81.
108   Ibid.
109   'Cigarette TV Ban – and It's Just the First Step', 1.
110   'Shock Plan to Ban Smoking in Public', 1.
111   Quoted after Taylor, *Smoke Ring*, 84.
112   Royal College of Physicians, *Smoking and Health Now*.
113   'Cigarettes on the Counter'.
114   Ibid.; 'Is Consumption Falling Off?'.
115   'A King-size Boom in Filter Tips', 15; 'One Man's Smoke'.
116   'Budget Accounts: On Schedule'; 'Imperial Tobacco'.
117   'The Man Who Worried: Suicide after the "Smokes" Warning', 2.
118   'Lulling the Smoker'.
119   Royal College of Physicians, *Smoking and Health Now*, 11.
120   'One Man's Smoke'.
121   'Consumption is Rising'.
122   'Plain or Filter?'.
123   Cartwright et al., 'Efficacy of an Anti-Smoking Campaign', 327.
124   Ibid.
125   Ibid., 328.
126   Ibid.
127   Cartwright et al., 'Health Hazards of Cigarette Smoking', 161.
128   'Liability to Lung Cancer: Incidence of Disease "Frightening"', 5.
129   Cartwright and Martin, 'Some Popular Beliefs Concerning the Causes of Cancer'.
130   Ibid.
131   Cartwright et al., 'Health Hazards of Cigarette Smoking', 166.
132   Cartwright et al., 'Efficacy of an Anti-Smoking Campaign', 328.
133   Jeger, 'The Social Implications', 94.
134   Ibid.
135   Ibid., 95.
136   Ibid.
137   Cartwright and Martin, 'Some Popular Beliefs Concerning the Causes of Cancer', 593.
138   Royal College of Physicians, *Smoking and Health*, 4.
139   Pyke, 'Cigarette Smoking and Bronchial Carcinoma'.
140   'Cigarettes in School Tuck Boxes'.
141   Cartwright et al., 'Distribution and Development of Smoking Habits'. In fact, the average amount of wine and spirit consumed per head in Britain had increased more sharply since the 1930s than cigarette consumption: 'Drinks and Smokes'.
142   Jeger, 'The Social Implications', 76.
143   Cole, 'The Economic Effects', 45.
144   Hooper, *Smoking Issues*, 3.
145   Doll, 'Occupational Lung Cancer'.
146   Ibid.; Uhlig, 'Über den Schneeberger Lungenkrebs'. There are clear links in these diagnoses to the early history of lung cancer, as discussed in Chapter 2, as a form of consumption that was not tuberculous.
147   Schüttmann, 'Beitrag zur Geschichte der Schneeberger Lungenkrankheit'; Schüttmann, 'Schneeberg Lung Disease and Uranium Mining in the Saxon Ore Mountains'; Lorenz, 'Radioactivity and Lung Cancer'.

148   Tweedale, 'Asbestos and Its Lethal Legacy'; Tweedale, *Magic Mineral to Killer Dust.*
149   Cole, 'The Economic Effects', 74.

Chapter 5   Trials and Tribulations: Lung Cancer Treatment, circa 1950 to 1970

1   Wheeler-Bennett, *King George VI*; Bradford, *King George VI.*
2   Quoted after Wheeler-Bennett, *King George VI*, 785.
3   Quoted after ibid.
4   Ibid., 787.
5   C. D., J. G. S. and A. T. J., 'Sir Clement Price Thomas (obituary)'.
6   Wheeler-Bennett, *King George VI*, 788.
7   Quoted after ibid.
8   Quoted after ibid., 789.
9   Quoted after ibid.
10   Quoted after ibid.
11   Ibid., 801.
12   Ibid., 802.
13   Harnett, *A Survey of Cancer in London*, 119.
14   Brooks et al., 'Carcinoma of the Bronchus'.
15   Smithers, 'Facts and Fancies About Cancer of the Lung'.
16   Gray, 'Sputum Cytodiagnosis in Bronchial Carcinoma'.
17   Smithers, 'Facts and Fancies About Cancer of the Lung'.
18   Harnett, *A Survey of Cancer in London*, 119.
19   Mackenzie, *Cancer*. Mackenzie, the second Baron Amulree, was a physician then working for the Ministry of Health; later he was a leading advocate of geriatric medicine in the UK.
20   Ibid., 16.
21   Ibid., 17.
22   Tod, *An Inquiry into the Extent to Which Cancer Patients in Great Britain Receive Radiotherapy*, 21.
23   Smithers, 'Facts and Fancies About Cancer of the Lung', 1237.
24   Ibid.
25   Davidson and Rouvray, *The Brompton Hospital*, 131–132.
26   Brooks et al., 'Carcinoma of the Bronchus'.
27   Radiotherapists pioneered this systematic, analytical approach to medical service provision in the interwar period: Pickstone, 'Contested Cumulations'.
28   Bignall et al., 'Survival in 6086 Cases of Bronchial Carcinoma'.
29   Crofton, 'John Guyett Scadding (obituary)'.
30   Valier and Timmermann, 'Clinical Trials'; Yoshioka, 'Streptomycin in Postwar Britain'; Yoshioka, 'Use of Randomisation in the Medical Research Council's Clinical Trial of Streptomycin'.
31   'Investigators Took Their Own Advice', 4.
32   Bignall, 'John Reginald Bignall (obituary)'.
33   Bignall, *Carcinoma of the Lung.*
34   Ibid., 204.
35   Ibid., 172–173.

36   This corresponds well with the data compiled by Harnett, *A Survey of Cancer in London*.

37   Bignall et al., 'Survival in 6086 Cases of Bronchial Carcinoma', 1067.

38   Much of the material in this section has been published previously in Timmermann, 'As Depressing as It Was Predictable'.

39   Medical Research Council, 'Medical Research Council's Statement on Tobacco Smoking and Cancer of the Lung'.

40   'Evaluation of Different Methods of Cancer Therapy – Recommendations of the Council's Steering Committee', FD 7/327, UK NA.

41   'Working Party for the Evaluation of Different Methods of Therapy in Carcinoma of the Bronchus, Minutes of the Meeting Held on 24 June 1958', FD 7/327, UK NA.

42   'Radiotherapy and Bronchial Carcinoma (editorial)'; Blanshard, 'The Palliation of Bronchial Carcinoma by Radiotherapy'; 'Radiotherapy for Lung Cancer (editorial)'.

43   On the history of radiotherapy in Britain, see Murphy, 'A History of Radiotherapy to 1950'; Cantor, 'The Definition of Radiobiology'.

44   Bradford Hill was also the statistician on most working parties, while other members varied.

45   'Evaluation of Different Methods of Cancer Therapy – Recommendations of the Council's Steering Committee'.

46   'Typescript: Evaluation of Different Methods of Cancer Therapy Committee', FD 7/340, UK NA.

47   Yoshioka, 'Streptomycin in Postwar Britain'.

48   On the history of the ICRF, see Austoker, *A History of the Imperial Cancer Research Fund*.

49   Valier and Timmermann, 'Clinical Trials'. On clinical research and the MRC, see also Booth, 'From Art to Science'.

50   See also Chalmers and Clarke, 'J Guy Scadding and the Move from Alternation to Randomization'.

51   'Working Party for the Evaluation of Different Methods of Therapy in Carcinoma of the Bronchus, Minutes of the Meeting Held on 23 June 1959', FD 7/327, UK NA.

52   Cf. Keating and Cambrosio, *Cancer on Trial*; Krueger, *Hope and Suffering*.

53   'Steering Committee for the Evaluation of Different Methods of Cancer Therapy, Minutes of the Meeting Held on 13 January 1958', FD 7/327, UK NA.

54   Ibid.

55   Ibid.

56   Ibid.

57   'Working Party for the Evaluation of Different Methods of Therapy in Carcinoma of the Bronchus, Notes for Discussion', FD 7/327, UK NA.

58   'Working Party for the Evaluation of Different Methods of Therapy in Carcinoma of the Bronchus, Minutes of the Meeting Held on 24 June 1958', FD 7/327, UK NA. See also Morrison et al., 'The Treatment of Carcinoma of the Bronchus'.

59   'Working Party for the Evaluation of Different Methods of Therapy in Carcinoma of the Bronchus, Minutes of the Meeting Held on 24 June 1958'. Scadding argued that it was not certain if the groups had been

strictly comparable in terms of operability. Also, often the histological diagnosis was only available after thoracotomy, and in such cases, although potentially responding well to radiotherapy, patients could not be included in a randomly allocated series.

60 Ibid.
61 Ibid.
62 Ibid.
63 Ibid. I will return to the issue of chemotherapy in the next chapter.
64 'Working Party for the Evaluation of Different Methods of Therapy in Carcinoma of the Bronchus, Memorandum', FD 7/327, UK NA.
65 D'Arcy Hart, 'Letter to Margaret Gorill, MRC', FD 23/1163, UK NA.
66 'Minutes of a Special Meeting with Consultant Radiotherapists on 21 January 1961'.
67 Ibid.
68 Ibid.
69 'Working Party for the Evaluation of Different Methods of Therapy in Carcinoma of the Bronchus, Memorandum [Draft]', FD 7/327, UK NA.
70 Ibid.
71 'Minutes of a Special Meeting with Consultant Radiotherapists on 21 January 1961', FD 7/327, UK NA.
72 Ibid.
73 Ibid.
74 Ibid.; see also Smart and Hilton, 'Radiotherapy of Cancer of the Lung'. On Hilton, see 'Obituary Notices: Gwen Hilton'.
75 'Minutes of a Special Meeting with Consultant Radiotherapists on 21 January 1961'.
76 Another option considered and later apparently dropped was a trial in fractionization of doses. See ibid.; and 'Working Party for the Evaluation of Different Methods of Therapy in Carcinoma of the Bronchus, Memorandum [Draft]'.
77 'Minutes of a Special Meeting with Consultant Radiotherapists on 21 January 1961'.
78 Gruhn, 'A History of the Histopathology of Lung Cancer'.
79 Watson and Berg, 'Oat Cell Lung Cancer'.
80 For results of some very early trials with nitrogen mustard, see Karnofsky et al., 'The Use of the Nitrogen Mustards in the Palliative Treatment of Carcinoma'.
81 Scadding et al., 'Comparative Trial of Surgery and Radiotherapy for the Primary Treatment of Small-Celled or Oat-Celled Carcinoma of the Bronchus'.
82 Ibid., 984.
83 Ibid., 985. Emphasis original.
84 Herrald, 'Handwritten Note', FD 7/1151, UK NA.
85 In fact, while originally assigned to the surgery group, he had become too breathless to withstand an operation and received palliative radio-therapy instead (and he was not the only member of this group who turned out to be inoperable when surgery was scheduled).
86 Miller et al., 'Five-year Follow-up of the Medical Research Council Comparative Trial of Surgery and Radiotherapy for the Primary

Treatment of Small-Celled or Oat-Celled Carcinoma of the Bronchus'; Fox and Scadding, 'Medical Research Council Comparative Trial of Surgery and Radiotherapy for Primary Treatment of Small-Celled or Oat-Celled Carcinoma of Bronchus: Ten-year Follow-up'.

87   Abbey Smith, 'Treatment of Bronchial Carcinoma (letter)'.
88   Belcher, 'Treatment of Bronchial Carcinoma (letter)'.
89   Harrison, 'Treatment of Bronchial Carcinoma (letter)'.
90   Meyer, 'Surgical Resection as an Adjunct to Chemotherapy for Small Cell Carcinoma of the Lung'.
91   Scadding, 'Treatment of Bronchial Carcinoma (letter)'.
92   FD 7/327 and FD 23/1163, UK National Archives.
93   For critical remarks on studies undertaken with chemotherapy in lung cancer from France, the US and Denmark, see Israel, 'Chemotherapy in Inoperable Bronchial Carcinoma (letter)'; Muggia et al., 'Treatment of Small-Cell Carcinoma of Bronchus (letter)'.
94   Medical Research Council Working Party, 'Study of Cytotoxic Chemotherapy as an Adjuvant to Surgery in Carcinoma of the Bronchus', 427.
95   Stott et al., '5-year Follow-up of Cytotoxic Chemotherapy'.
96   'Section of Radiology: Discussion on the Place of Miniature Radiography in the Diagnosis of Diseases of the Chest'.
97   Ellman, *Essentials in Diseases of the Chest for Students and Practitioners*, 46.
98   Posner et al., 'Mass Radiography and Cancer of the Lung'. The authors did not use the language of risk to which we have since become accustomed.
99   Ibid.
100   Ibid.; Posner et al., 'Place of Mass Radiography in Relation to Lung Cancer'.
101   Cf. 'Early Diagnosis of Lung Cancer (editorial)'.
102   Posner et al., 'Place of Mass Radiography in Relation to Lung Cancer'.
103   This detection rate (0.4 per thousand) was the same as that for the country as a whole but lower than the rates achieved in intensive mass radiography campaigns in Glasgow, Edinburgh and Liverpool, where up to 70 percent of the male population were screened within a few weeks.
104   Posner et al., 'Place of Mass Radiography in Relation to Lung Cancer', 1159. I presume the calculation is based on the assumption that screening was done exclusively for one or the other disease. If more than one disease could be detected by the same screening test, the test was accordingly more cost effective. On the costs of screening, see also Commission on Chronic Illness, *Prevention of Chronic Illness*, 2, 50–51.
105   Posner et al., 'Place of Mass Radiography in Relation to Lung Cancer'.
106   Barrett, 'Introduction'.
107   Ibid.
108   Ibid., 1–2.
109   Ibid., 2.
110   Ibid.
111   Ibid., 4.
112   Allison and Temple, 'The Future of Thoracic Surgery'.
113   Ibid.

114 Belcher, 'Indications for Surgery and Choice of Operation'.
115 Simon, 'Radiographic Aspects of Carcinoma of the Lung', 87.
116 Deeley, 'The Early Diagnosis of Lung Cancer'.
117 Bignall, 'Early Diagnosis of Bronchial Carcinoma'.
118 Barrett, for example, started his address with the following words: 'It is my function to introduce the subject of carcinoma of the bronchus in a general way, and I am well chosen for this because I am a surgeon.' Barrett, 'Introduction', 1.
119 Pack and Ariel, 'A Half Century of Effort to Control Cancer'.
120 Ikeda, 'Flexible Bronchofibrescope'; Mitchell et al., 'Fibreoptic Bronchoscopy: Ten Years on (Occasional Review)'.
121 Thatcher and Spiro, *New Perspectives in Lung Cancer*.
122 Smart and Hilton, 'Radiotherapy of Cancer of the Lung'. This seems to support Belcher's fatalistic notion that it is the 'the disease process itself' that decides over the outcome of therapy, more than the modality chosen.
123 Contribution by Smart to the discussion following Jones, 'Megavoltage Radiotherapy of Carcinoma of the Bronchus', 97.
124 Belcher, 'Indications for Surgery and Choice of Operation', 100.
125 One of these colleagues was Roger Abbey Smith, who wrote a useful insider history of lung surgery in Britain: Abbey Smith, 'Development of Lung Surgery'.
126 On another club of this type, see Milstein, 'The Cardiothoracic Society (Pete's Club) 1959 to 1989'. The minute books of Charlie's Club can be viewed in the Archives of the Royal College of Surgeons in London.
127 In 1983 Belcher commented that: 'As bronchial carcinoma is so common and surgery offers almost the only hope of prolonged survival, it is surprising that so few large series have been reported.' Belcher, 'Thirty Years of Surgery for Carcinoma of the Bronchus', 430.
128 It did make a difference for operative mortality, though. More patients died after pneumonectomies.
129 Belcher, 'Thirty Years of Surgery for Carcinoma of the Bronchus'.
130 Harnett, *A Survey of Cancer in London*, 119.

Chapter 6   More Enthusiasm, Please: Preventing, Screening, Treating, Classifying, circa 1960 to 1990

1 Royal College of Physicians, *Smoking and Health Now*.
2 Keating and Cambrosio, *Cancer on Trial*; Krueger, *Hope and Suffering*.
3 Berridge, *Marketing Health*.
4 Keating and Cambrosio, *Cancer on Trial*.
5 Holmes Sellors, 'The Management of Cancer of the Lung'.
6 Meyer, 'Surgical Resection as an Adjunct to Chemotherapy for Small Cell Carcinoma of the Lung'.
7 Holmes Sellors, 'The Management of Cancer of the Lung'.
8 'Minutes of a Meeting to Consider the Co-ordination of Clinical Trials in Lung Cancer, Held at MRC Head Office on 3 November 1982', FD 7/3068, UK NA.

9   Holmes Sellors, 'The Management of Cancer of the Lung', 20.
10 Ibid., 22.
11 Ibid.
12 Ibid.
13 Wilkes, 'Terminal Cancer at Home'. On the history of the hospice and palliative care movement, see Clark et al., *A Bit of Heaven for the Few?*; Clark, 'From Margins to Centre'; Clark, 'Cradled to the Grave?'.
14 Twycross, 'The Terminal Care of Patients with Lung Cancer'.
15 Wilkes, 'Terminal Cancer at Home', 800.
16 Ibid.
17 Clark et al., *A Bit of Heaven for the Few?*.
18 Marie Curie Memorial Foundation, *The Marie Curie Memorial Foundation*.
19 Joint National Cancer Survey Committee of the Marie Curie Memorial and the Queen's Institute of District Nursing, *Report on a National Survey Concerning Patients with Cancer Nursed at Home*. While cases in the 1952 sample are not classified according to the site of cancer, it is clear that, based on the symptoms, lung cancer cases form a small minority. However, as lung cancer patients were usually men, the majority would be cared for by their wives and thus not included in the survey. Nearly twice as many women were included as men, presumably suffering from breast and gynaecological cancers, followed by patients with symptoms pointing towards cancers of the digestive tract.
20 Rossi, *Fighting Cancer with more than Medicine*.
21 Craig, *The Last Freedom* provides an example for the important role of the Macmillan nurse for making the final months of a lung cancer patient in the UK more bearable in the mid 1990s.
22 Berridge, *Marketing Health*.
23 'Attitudes to Smoking (editorial)'.
24 Hirayama, 'Non-smoking Wives of Heavy Smokers Have a Higher Risk of Lung Cancer'. Cf. Berridge, 'Passive Smoking and Its Prehistory in Britain'.
25 George Godber, 'Minute', 18 June 1969, MH 154/169, UK National Archives, cited after Berridge, *Marketing Health*, 169. On the US, see Brandt, *The Cigarette Century*.
26 Berridge, *Marketing Health*, 168; Webster, 'Tobacco Smoking Addiction'. On the MRC unit, see Murphy, 'The Early Days of the MRC Social Medicine Research Unit'. On Ball, see Kirby and Richmond, 'Obituary: Keith Ball'.
27 Daube in an interview with the journalist William Norman, R. 12, Box 77, SA/ASH, Wellcome Library, quoted after Berridge, *Marketing Health*, 173.
28 Ibid.
29 Ibid., 174.
30 Taylor, *Smoke Ring*.
31 On the de-stigmatization of breast cancer, see King, *Pink Ribbons, Inc.*
32 Morabia and Zhang, 'History of Medical Screening'.
33 For a recent, critical assessment of cancer screening, Welch, *Should I Be Tested for Cancer?*.

34 Aronowitz, 'Do Not Delay'; see also Aronowitz, *Unnatural History*, chap. 6.
35 Morabia and Zhang, 'History of Medical Screening'.
36 Löwy, *Preventive Strikes*; Gardner, *Early Detection*; Vayena, 'Cancer Detectors'.
37 Morabia and Zhang, 'History of Medical Screening', 467.
38 Löwy, *Preventive Strikes*.
39 Shapiro et al., *Periodic Screening for Breast Cancer*.
40 Welch, *Should I Be Tested for Cancer?*, 154–155.
41 Huggins et al., 'Edinburgh Trial of Screening for Breast Cancer'.
42 Welch, *Should I Be Tested for Cancer?*, 155.
43 Advisory Committee on Breast Cancer Screening, *Screening for Breast Cancer in England*.
44 Lister, 'NHS Accused over Women's Breast Cancer Screening Risks'.
45 Miller et al., 'Canadian National Breast Screening Study'.
46 Olsen and Gøtzsche, 'Cochrane Review on Screening for Breast Cancer with Mammography'; Kolata, 'Study Sets Off Debate Over Mammograms' Value'.
47 On a catalogue of problems surrounding the evaluation of screening tests, see Welch, *Should I Be Tested for Cancer?*, 152–176.
48 'Section of Radiology: Discussion on the Place of Miniature Radiography in the Diagnosis of Diseases of the Chest'.
49 Posner et al., 'Place of Mass Radiography in Relation to Lung Cancer'. See Chapter 5.
50 Nash et al., 'South London Lung Cancer Study'; 'Early Diagnosis of Lung Cancer (editorial)'.
51 Nash et al., 'South London Lung Cancer Study'; 'Early Diagnosis of Lung Cancer (editorial)'; see also Kubik, 'Screening for Lung Cancer High-risk Groups (letter)'.
52 Brett, 'Earlier Diagnosis and Survival in Lung Cancer'.
53 'Folder: Influence of Delay in Diagnosis on the Prognosis of Cancer: Reports and Minutes of Meetings', MH 160/675, UK NA.
54 Doll, 'Letter to Pamela Aylett', MH 160/675, UK NA.
55 'Screening for Cancer', 4.
56 Ibid.
57 Ibid.
58 Ministry of Health – Central Health Services Council – Standing Tuberculosis Advisory Committee, *The Future of the Chest Services*, 3.
59 Figures: Ibid.
60 'Mass Miniature Radiography Service, House of Commons Debate, 2 March 1971'.
61 'X-ray Units, House of Commons Debate, 30 Jan 1978'.
62 Figures from a written answer by John Dunwoody, Secretary of State for Social Services, 'Mass X-ray Service, House of Commons Debate, 11 May 1970'.
63 Ellis and Gleeson, 'Lung Cancer Screening'.
64 The Johns Hopkins (10,387 volunteers enrolled between 1973 and 1978) and Memorial Sloan Kettering studies (10,040 volunteers enrolled

between 1974 and 1978) were similar in design. Fontana et al., 'Screening for Lung Cancer'.

65 Ellis and Gleeson, 'Lung Cancer Screening'.

66 Ibid.

67 Meyer, 'Growth Rate Versus Prognosis in Resected Primary Bronchogenic Carcinomas'.

68 Cf. Aisner, 'CT Screening for Lung Cancer'.

69 On the emergence of chemotherapy and its relationship with the older treatment modalities, see Pickstone, 'Contested Cumulations'.

70 Keating and Cambrosio, *Cancer on Trial*.

71 Herrald, 'Memo to Dr Gray', FD 7/340, UK NA.

72 Neale, 'Note for File'; Neale, 'Memo to Dr Norton', n.d.; see also Neale, 'Letter to Brian Windeyer', 1 December 1969 all FD 7/340, UK NA.

73 Neale, 'Letter to Brian Windeyer', 30 November 1972, FD 7/340, UK NA.

74 Neale, 'Memo to Dr Norton', n.d., FD 7/340, UK NA.

75 Ibid.

76 Neale, 'Memo to Dr Norton', 15 September 1971, FD 7/340, UK NA.

77 Neale, 'Memo to Dr Norton', 28 January 1972, FD 7/340, UK NA.

78 Ibid.

79 The proposed members, as of January 1973, were Prof. N. M. Bleehen (Chairman), Dr T. A. Connors, Dr T. J. Deeley, Dr S. Dische, Prof. W. Duncan, Prof. G. Hamilton Fairley, Dr J. F. Fowler, Dr K. E. Halnan, Prof. L. F. Lamerton, Mr R. Peto, Dr M. C. Pike, Dr G. Wiernik, Prof. R. Sealy (corresponding member), Dr I. Todd, Prof. R. J. Walton (corresponding member), and Dr E. J. E. Neale (Secretary). See 'Annex 2 to Circulation to Council, MRC 73/10: Radiotherapy Working Party', FD 7/340, UK NA.

80 'Extract from Council Minutes for Meeting of Jan 1973', FD 7/340, UK NA.

81 Halnan, 'Fifty Years of the National Health Service'.

82 Cf. Bleehen, 'Chemotherapy for Carcinoma of the Bronchus'; Carmichael, 'Introduction'.

83 Cf. Pickstone, 'Contested Cumulations'.

84 On the treatment of lung cancer in the 1980s, see Smyth, *The Management of Lung Cancer*; Bates, *Bronchial Carcinoma*.

85 Cf. Pickstone, 'Contested Cumulations'; see also Halnan, 'Fifty Years of the National Health Service'.

86 Honess, 'Obituary: Norman Bleehen'.

87 Cf. Ueyama and Lecuyer, 'Building Science-based Medicine at Stanford'.

88 Honess, 'Obituary: Norman Bleehen', 681.

89 'Obituary: Professor Norman Bleehen 1930–2008'.

90 Honess, 'Obituary: Norman Bleehen'; Bleehen, 'Obituary: Norman Montague Bleehen'; 'MRC Clinical Trials Unit: Unit Profile'.

91 Bleehen, 'Chemotherapy for Carcinoma of the Bronchus'.

92 'Lung Cancer Working Group, Minutes of the Meeting Held on 23 February 1973', FD 7/383, UK NA.

93 'Lung Cancer Working Group, Minutes of the Meeting Held on 16 July 1973', FD 7/383, UK NA. See also Laing et al., 'Treatment of Small-Cell Carcinoma of Bronchus'.

94 Evaluation of Cancer Therapy Committee, 'Folder: Radiotherapy Working Party – Information on Current Clinical Trials', FD 7/366, UK NA.
95 Pett et al., 'Lung Cancer'. See also Thatcher et al., 'Chemotherapy in Non-small Cell Lung Cancer'.
96 Austoker, *A History of the Imperial Cancer Research Fund*, 306–315; Pickstone, 'Contested Cumulations'.
97 Interview with Professor Nick Thatcher, medical oncologist, Christie Hospital, Manchester.
98 Thatcher et al., 'Chemotherapy in Non-small Cell Lung Cancer'.
99 Ibid., S84.
100 Ibid., S85.
101 Ibid., S94.
102 Ibid.
103 Ibid., S95.
104 Interview with Professor Tom Treasure, thoracic surgeon, Guy's Hospital, London.
105 Mountain, 'A New International Staging System for Lung Cancer'.
106 Milroy, 'Staging of Lung Cancer', 593.
107 Mountain et al., 'A System for the Clinical Staging of Lung Cancer'.
108 The official journal of the International Association for the Study of Lung Cancer is since 2006 the *Journal of Thoracic Oncology*. Cf. Jett, 'The New Official Journal of the IASLC'.
109 World Health Organization, *Histological Typing of Lung Tumours*.
110 World Health Organization, *Manual of the International Statistical Classification of Diseases, Injuries and Causes of Death*.
111 Cf. World Health Organization, 'General Preface to the Series'; Sobin, 'International Histological Classification of Tumors, Second Edition'.
112 Goldstraw and Crowley, 'The International Association for the Study of Lung Cancer International Staging Project on Lung Cancer'.
113 Menoret, 'The Genesis of the Notion of Stages in Oncology'.
114 Doll, 'The Pierre Denoix Memorial Lecture'.
115 Copeland, 'American Joint Committee on Cancer Staging and End Results Reporting'.
116 Carr, 'A Report on the Development of the Staging System for Cancer of the Lung'.
117 Mountain et al., 'A System for the Clinical Staging of Lung Cancer'.
118 Carr, 'Classification of Lung Cancer'; Sobin and Baker, 'Request for Suggestions on a Unified System for Classifying and Staging Cancer'.
119 Mountain, 'A New International Staging System for Lung Cancer'; see also Mountain, 'Staging of Lung Cancer'.
120 Spiro, 'Diagnosis and Staging'; Golding, 'The Role of Computer Tomography in the Management of Bronchial Carcinoma'. On controversies related to scanning, see also Pett et al., 'Lung Cancer'.
121 Interview with Professor Peter Goldstraw, thoracic surgeon, Brompton Hospital, London.
122 Ibid.
123 Ibid.
124 Ibid.

125 Ibid.
126 Hansen, *History of the IASLC 1972–2007.*
127 'International Association for the Study of Lung Cancer'.
128 Ibid.; and Hansen, 'International Association for the Study of Lung Cancer'.
129 Jett, 'The New Official Journal of the IASLC'.
130 Goldstraw and Crowley, 'The International Association for the Study of Lung Cancer International Staging Project on Lung Cancer'.
131 Pett et al., 'Lung Cancer'.
132 Ibid.
133 Ibid.
134 Co-ordinating Committee on Cancer Research, 'Folder: Sub-Committee for the Co-ordination of Clinical Trials in Lung Cancer. Setting up/Membership + First Meeting', FD 7/3068, UK NA.
135 Cf. Austoker, *A History of the Imperial Cancer Research Fund,* 306–315; Pickstone, 'Contested Cumulations'.
136 Smyth, 'Letter to Sir Michael Stoker', FD 7/3068, UK NA.
137 'Minutes of a Meeting to Consider the Co-ordination of Clinical Trials in Lung Cancer, Held at MRC Head Office on 3 November 1982', FD 7/3068, UK NA.
138 'Draft Minutes [handwritten, Probably for the First Meeting of the Subcommittee on 10 January 1985]', FD 7/3068, UK NA.
139 Ibid.
140 Ibid.
141 Ibid.
142 'Comments Received on Returned Questionnaires', FD 7/3075, UK NA.
143 Ibid.

Chapter 7 The Management of Stigma: Lung Cancer and Charity, circa 1990 to 2000

1 Percentage figures from Samet et al., 'Lung Cancer in Never Smokers'. Samet and his co-authors deplore that 'given the impact of this disease, there is surprisingly little information available on the descriptive epidemiology of lung cancer in never smokers' and discuss some of the reasons: cancer registries or death certificates, for example, rarely provide reliable information on smoking histories, and cohort studies have not always been helpful either: ibid., p. 5626. The total incidence of all different types of leukaemia in the United Kingdom in 2006, according to figures compiled by Cancer Research UK, was 7,237 – not very much higher. Cf. Cancer Research UK, 'CancerStats Key Facts'.
2 Goffman, *Stigma,* 11.
3 See, for example, Herek, 'AIDS and Stigma'.
4 Bouluri, 'Dundee Mum Tells of Lung Cancer Ordeal'.
5 Ibid.
6 Goddard, 'The Winston Man Dies of Lung Cancer'. Lack of awareness, indeed, as we have also seen in Chapter 4, is not the problem.

7   Comment: 'Libby, London, UK, 04/3/2009 07:51', ibid., accessed on 24 May 2010.
8 .   Chapple et al., 'Stigma, Shame, and Blame Experienced by Patients with Lung Cancer'.
9   Henschke and McCarthy, *Lung Cancer*, 13.
10   Ibid.
11   Interview with Professor Nick Thatcher, medical oncologist, Christie Hospital, Manchester.
12   Kmietowicz, 'Research Spending on Cancer Doesn't Match Their Death Rates'.
13   Parker-Pope, 'Cancer Funding'.
14   Burnet et al., 'Years of Life Lost (YLL) from Cancer is an Important Measure of Population Burden'.
15   Statistics are flexible, though: Yet another way of measuring the impact of a malignant disease, Burnet and colleagues suggested, were average years of life lost per affected patient (AYLL), representing individual cancer burdens. When they plotted AYLL against NCRI funding, they found that the cancers which appeared to be underfunded, were above all brain and CNS, along with cervical, melanoma and kidney cancer – diseases that tend to affect younger patients. Ibid.
16   Peters, 'Should Smokers Be Refused Surgery? Yes'.
17   Glantz, 'Should Smokers Be Refused Surgery? No'.
18   Elliot, 'Growing up and Giving Up'.
19   Viscusi, *Smoking: Making the Risky Decision*, 77–78.
20   Conrad et al., 'Why Children Start Smoking Cigarettes'.
21   Stapleton, 'Cigarette Smoking Prevalence, Cessation and Relapse'.
22   Cf. Eysenck, *Smoking, Health and Personality*; Fisher, *Smoking: The Cancer Controversy*. See also Buchanan, *Playing with Fire*; Baines, 'Cancer and the Individual in Britain', 155–159.
23   Eysenck et al., 'Smoking and Personality'.
24   Bahnson, 'In Memory of Dr David M. Kissen'.
25   'Obituary Notices: D. M. Kissen'.
26   Quoted after Bahnson, 'In Memory of Dr David M. Kissen', 313.
27   See also Payne, '"Smoke Like Man, Die Like a Man"?'.
28   Todd, *Social Class Variations in Cigarette Smoking and Mortality from Associated Diseases*.
29   Ibid., 26.
30   Department of Health, *Statistics on Smoking*.
31   'Smokers' Kids "Are Yobs"'. The story was widely reported. See also, for example, 'Teen Behaviour "Better Than 1985"'.
32   Cox and Pritchard, 'Smoking Parents'.
33   Palmer, 'Smoking, Caning, and Delinquency in a Secondary Modern School'.
34   Elliot, 'Growing Up and Giving Up'.
35   Cox and Pritchard, 'Smoking Parents'. See also Bournemouth University Press Office, 'Kids Worse Off with Smoking Parents'.
36   For Britain, see Berridge, *Marketing Health*. See also Snowdon, *Velvet Glove, Iron Fist*. For the activists' perspective on the US campaigns, see

Glantz and Balbach, *Tobacco War*; and for a critical account: Sullum, *For Your Own Good*.

37   Markle and Troyer, 'Smoke Gets in Your Eyes', 617.

38   Snowdon, *Velvet Glove, Iron Fist*, 213–268.

39   Ibid., 245.

40   Cruickshanks et al., 'Cigarette Smoking and Hearing Loss'.

41   Cf. Kabat, *Hyping Health Risks*, 147–182.

42   Snowdon, *Velvet Glove, Iron Fist*, 248.

43   Cf. Glantz and Balbach, *Tobacco War*.

44   Bayer and Stuber, 'Tobacco Control, Stigma, and Public Health'.

45   Ibid., 49.

46   Barnes, 'Caring and Curing'; Barnes, 'Captain Chemo and Mr Wiggly'.

47   See, for example, King, *Pink Ribbons, Inc.*; Ehrenreich, 'Smile! You've Got Cancer'.

48   On changing perceptions of breast cancer in Britain, see Baines, 'Three Stories'.

49   Cf. Gaudillière, 'The Molecularization of Cancer Etiology'; Löwy, 'Cancer'.

50   The history of the ICRF has been well documented by Austoker, *A History of the Imperial Cancer Research Fund*. Ross, *Crusade*.

51   Gaudillière, 'The Molecularization of Cancer Etiology'.

52   See, for example, Sporn, 'The War on Cancer'.

53   The Christie Hospital, for example, is part of the fabric of Manchester. Most Mancunians know somebody who has received treatment at the Christie, and many have donated to the charity. On the Christie, see Magnello, *A Centenary History of the Christie Hospital*. On the history of charity hospitals, see also Prochaska, *Philanthropy and the Hospitals of London*.

54   Rossi, *Fighting Cancer with More Than Medicine*; Clark, 'From Margins to Centre'.

55   Donnelly, *Cinderella Cancer*.

56   A similar argument is made by Henschke and McCarthy, *Lung Cancer*.

57   As Malcolm Nicolson and George Lowis point out in an article on the history of the British Multiple Sclerosis Society in 2002, little work has done on the history of such single disease charities, in spite of their growing public visibility: Nicolson and Lowis, 'The Early History of the Multiple Sclerosis Society of Great Britain and Northern Ireland'.

58   Donnelly, *Cinderella Cancer*, 10.

59   Ibid.

60   Story as told by Donnelly himself: Ibid., 11–14; Interview with Professor Ray Donnelly, thoracic surgeon, Roy Castle Foundation, Liverpool.

61   Donnelly, *Cinderella Cancer*, 22–26. See also 'Landmark for Keyhole Surgery'.

62   Donnelly, *Cinderella Cancer*, 66, 71, 81.

63   Scouse is colloquial for the dialect spoken in Liverpool and parts of Merseyside.

64   The words 'You'll Never Walk Alone' also feature in the Liverpool F.C. crest and on one of the iron gates to the club's Anfield stadium.

65   Murden, 'City of Change and Challenge'.

66 The *Sun* newspaper is still boycotted widely in Liverpool because of its defamatory headlines after a crowd crush killed 96 Liverpool fans at Hillsborough stadium, home of Sheffield Wednesday Football Club, on 15 April 1989.

67 Donnelly, *Cinderella Cancer*, 63.

68 Hebblethwaite, 'Obituary: The Most Rev Derek Worlock'.

69 Donnelly, *Cinderella Cancer*, 39.

70 Ibid.

71 Holmes Sellors, 'The Management of Cancer of the Lung', 19.

72 Cf. 'Homepage: Dimbleby Cancer Care'; 'Obituary: Mr Richard Dimbleby: Professionalism in Broadcasting'; 'Homepage: Maggie's'; Powell, 'Obituary: Maggie Jencks'.

73 Clark et al., *A Bit of Heaven for the Few?*.

74 'Obituary: Roy Castle'.

75 Barker, 'Leave 'em Laughing: Obituary of Roy Castle'.

76 Castle, *Now and Then*, 187.

77 Ibid., 198.

78 Donnelly, *Cinderella Cancer*, 40.

79 Ibid.

80 Ibid., 58.

81 Ibid., 71–72.

82 Ibid., 56.

83 On the history of the Brompton Hospital, see Davidson and Rouvray, *The Brompton Hospital*.

84 On the history of these organizations, see Clark et al., *A Bit of Heaven for the Few?*, 89–98.

85 Donnelly, *Cinderella Cancer*, 32–33.

86 'Terry Kavanagh – Conquering Lung Cancer – I Can Get Well'.

87 Donnelly, *Cinderella Cancer*, 33.

88 Ibid., 34.

89 Ibid., 85.

90 Field and Youngson, 'The Liverpool Lung Project'. AstraZeneca, the manufacturers of tamoxifen, have also taken an active interest in and supported the research programme of the Roy Castle Foundation; see Donnelly, *Cinderella Cancer*, 139.

91 Kolata, 'Study Says Beetter Scans May Discover Lung Cancer Sooner'; Black and Baron, 'CT Screening for Lung Cancer: Spiraling into Confusion? [editorial]'; Welch et al., 'How Two Studies on Cancer Screening Led to Two Results'.

92 Field and Duffy, 'Lung Cancer Screening: The Way Forward'.

93 Henschke et al., 'Early Lung Cancer Action Project: Overall Design and Findings from Baseline Screening'; Henschke et al., 'Survival of Patients with Stage I Lung Cancer Detected on CT Screening', 1763.

94 Kolata, 'Study Says Beetter Scans May Discover Lung Cancer Sooner'.

95 Kolata, 'Researchers Dispute Benefits of CT Scans for Lung Cancer'; Welch et al., 'How Two Studies on Cancer Screening Led to Two Results'.

96 Woloshin et al., 'Tobacco Money: Up in Smoke?'.

97 NY-ELCAP investigators, 'Tobacco Money'.

98 Woloshin et al., 'Tobacco Money: Up in Smoke?'.

99   Harris, 'Cigarette Company Paid for Lung Cancer Study'.
100 Ibid.
101 Aberle et al., 'Reduced Lung-cancer Mortality with Low-dose Computed Tomographic Screening'.
102 Duke and Eisen, 'Finding Needles in a Haystack'.
103 Edey and Hansell, 'CT Lung Cancer Screening in the UK'.
104 Field, 'Lung Cancer Risk Models Come of Age'.
105 Field and Youngson, 'The Liverpool Lung Project'.
106 Field et al., 'The Liverpool Lung Project Research Protocol'.
107 See also Clark et al., *A Bit of Heaven for the Few?*, 68–69.
108 Herbert, 'Head of Cancer Fund Quits over Donations'.
109 Donnelly, *Cinderella Cancer*, 164–166.
110 See Epstein, *Impure Science*; Lerner, *The Breast Cancer Wars*; Lerner, *When Illness Goes Public: Celebrity Patients and How We Look at Medicine*.

Chapter 8   Still Recalcitrant? Some Conclusions

1 In the USA a peak in both incidence and mortality rates for men was reached in the mid 1980s. Since then both incidence and mortality has been declining in all age cohorts apart from the very old. For women, rates were levelling off overall after 2000 and pointing downwards for all under 65. See Jemal et al., 'Annual Report to the Nation on the Status of Cancer'; Spiro et al., 'Lung Cancer'.
2 Peake et al., *National Lung Cancer Audit Report 2011*.
3 Aronowitz, *Unnatural History*; Lerner, *The Breast Cancer Wars*.
4 Fox, 'On the Diagnosis and Treatment of Breast Cancer'. I am grateful to Ilana Löwy for pointing this out to me. On the history of breast cancer, see also Aronowitz, *Unnatural History*; Lerner, *The Breast Cancer Wars*.
5 Ehrenreich, 'Smile! You've Got Cancer'.
6 Abel and Subramanian, *After the Cure*; Aronowitz, 'The Converged Experience of Risk and Disease'.

Bibliography

Archival sources

Allison, P. R. 'Letter to Arthur Tudor Edwards', 26 September 1938. Bryce Papers, MS0005. Royal College of Surgeons Archives.

'Annex 2 to Circulation to Council, MRC 73/10: Radiotherapy Working Party', January 1973. FD 7/340. UK National Archives.

Brompton Hospital. 'Medical Report for 1906', 1906. BH/A/13. Royal London Hospital Archives.

————. 'Medical Report for 1914', 1914. BH/A/13. Royal London Hospital Archives.

————. 'Medical Report for 1920', 1920. BH/A/13. Royal London Hospital Archives.

————. 'Medical Report for 1921', 1921. BH/A/13. Royal London Hospital Archives.

————. 'Medical Report for 1922', 1922. BH/A/13. Royal London Hospital Archives.

Bryce, A. Graham. 'Letter to Hugh Morriston Davies', 2 November 1931. Bryce Papers, MS0005. Royal College of Surgeons Archives.

————. 'Letter to J. E. H. Roberts', 2 November 1931. Bryce Papers, MS0005. Royal College of Surgeons Archives.

————. 'Letter to J. E. H. Roberts', 9 May 1933. Bryce Papers, MS0005. Royal College of Surgeons Archives.

————. 'Letter to J. E. H. Roberts', 21 October 1941. Bryce Papers, MS0005. Royal College of Surgeons Archives.

Bundy. 'Memo to Kearney', 4 August 1943. BW 4/35. UK National Archives.

'Circular', 15 February 1939. Bryce Papers, MS0005, 20. Royal College of Surgeons Archives.

'Comments Received on Returned Questionnaires', n.d. FD 7/3075. UK National Archives.

Co-ordinating Committee on Cancer Research. 'Folder: Sub-Committee for the Co-Ordination of Clinical Trials in Lung Cancer. Setting up/Membership + First Meeting', 1985 1984. FD 7/3068. UK National Archives.

D'Arcy Hart, Philip. 'Letter to Margaret Gorill, MRC', 15 March 1960. FD 23/1163. UK National Archives.

Doll, Richard. 'Cancer of the Lung Investigation: Interim Report', 1 May 1948. FD 1/1989. UK National Archives.

————. 'Letter to Pamela Aylett', 1 May 1968. MH 160/675. UK National Archives.

'Dr Hugh Morriston Davies Conversation', n.d. Archives and Records Department, Bangor University.

'Draft Minutes' [handwritten, probably for the First Meeting of the Subcommittee on 10 January 1985], n.d. FD 7/3068. UK National Archives.

Elton, Arthur. 'Letter to A. F. Primrose (British Council)', 21 August 1943. BW 4/35. UK National Archives.

Evaluation of Cancer Therapy Committee. 'Folder: Radiotherapy Working Party – Information on Current Clinical Trials', 1971. FD 7/366. UK National Archives.

'Evaluation of Different Methods of Cancer Therapy – Recommendations of the Council's Steering Committee', 1957. FD 7/327. UK National Archives.

'Extract from Council Minutes for Meeting of Jan 1973', January 1973. FD 7/340. UK National Archives.

Farrow, R. J. R. 'Letter to J. E. Pater (Ministry of Health)', 28 May 1946. FD 1/1989. UK National Archives.

'Folder: Cancer of the Lung: Res. On, E. L. Kennaway', 1947–1948. FD 1/1990. UK National Archives.

'Folder: Film – Thoracic Surgery', n.d. BW 4/35. UK National Archives.

'Folder: Influence of Delay in Diagnosis on the Prognosis of Cancer: Reports and Minutes of Meetings', 1967–1969. MH 160/675. UK National Archives.

Herrald, J. R. 'Handwritten Note', 22 August 1966. FD 7/1151. UK National Archives.

Herrald, J. R. 'Memo to Dr Gray', 23 December 1969. FD 7/340. UK National Archives.

Hill, Austin Bradford. 'Letter to Edward Mellanby', 19 June 1947. FD 1/1993. UK National Archives.

———. 'Letter to Frank H. K. Green', 15 July 1946. FD 1/1989. UK National Archives.

Hill, Austin Bradford, E. L. Kennaway and Percy Stocks. 'Proposed Investigation of Cancer of the Lung', n.d. FD 1/1989. UK National Archives.

'Lung Cancer Working Group, Minutes of the Meeting Held on 16 July 1973', 1973. FD 7/383. UK National Archives.

'Lung Cancer Working Group, Minutes of the Meeting Held on 23 February 1973', 1973. FD 7/383. UK National Archives.

'Medical Research Council, Conference on Cancer of the Lung Held on 29 September 1947, Minutes', 1947. FD 1/1989. UK National Archives.

'Medical Research Council, Conference on Cancer of the Lung Held on 6 February 1947, Minutes', 1947. FD 1/1989. UK National Archives.

Mellanby, Edward. 'Letter to Austin Bradford Hill', 16 June 1947. FD 1/1993. UK National Archives.

———. 'Memorandum', 18 June 1946. FD 1/1989. UK National Archives.

'Memorandum by the Society of Thoracic Surgeons of Great Britain and Ireland', May 1952. BD18/901. UK National Archives.

'Minutes of a Meeting to Consider the Co-ordination of Clinical Trials in Lung Cancer, Held at MRC Head Office on 3 November 1982', 1982. FD 7/3068. UK National Archives.

'Minutes of a Special Meeting with Consultant Radiotherapists on 21 January 1961', 1961. FD 7/327. UK National Archives.

Neale, Julie. 'Letter to Brian Windeyer', 1 December 1969. FD 7/340. UK National Archives.

———. 'Letter to Brian Windeyer', 30 November 1972. FD 7/340. UK National Archives.

———. 'Memo to Dr Norton', 15 September 1971. FD 7/340. UK National Archives.

———. 'Memo to Dr Norton', 28 January 1972. FD 7/340. UK National Archives.

————. 'Memo to Dr Norton', n.d. FD 7/340. UK National Archives.

————. 'Note for File', 10 December 1969. FD 7/340. UK National Archives.

Nelson, Henry P. 'Letter to Alexander Graham Bryce', 3 December 1934. Bryce Papers, MS0005. Royal College of Surgeons Archives.

Primrose, A. F. 'Letter to Arthur Elton (Ministry of Information)', 25 August 1943. BW 4/35. UK National Archives.

Reynolds, J. S. 'To Mr I. McLeod, Minister of Health', 15 February 1954. MH 55/1012. UK National Archives.

'Screening for Cancer', n.d. MH 160/675. UK National Archives.

Smyth, John F. 'Letter to Sir Michael Stoker', 17 May 1984. FD 7/3068. UK National Archives.

Society of Thoracic Surgeons of Great Britain and Ireland. 'Memorandum for the Ministry of Health by the Society of Thoracic Surgeons of Great Britain and Ireland', May 1952. BD18/901. UK National Archives.

————. 'Memorandum on the Provision of a National Thoracic Surgery Service', November 1943. Bryce Papers, MS0005, 25. Royal College of Surgeons Archives.

————. 'Memorandum on the Provision of a National Thoracic Surgery Service', March 1948. Copy in the author's possession.

'Steering Committee for the Evaluation of Different Methods of Cancer Therapy, Minutes of the Meeting Held on 13 January 1958', 1958. FD 7/327. UK National Archives.

Stocks, Percy. 'Medical Research Council, Cancer of the Lung (Memorandum Prepared by Dr. Percy Stocks)', 22 November 1946. FD 1/1989. UK National Archives.

Todd, G. F. 'Typescript: Smoking and Lung Cancer: a Statistical Survey', n.d. FD 1/2009. UK National Archives.

'Typescript: Evaluation of Different Methods of Cancer Therapy Committee', n.d. FD 7/340. UK National Archives.

'Working Party for the Evaluation of Different Methods of Therapy in Carcinoma of the Bronchus, Memorandum', 8 October 1959. FD 7/327. UK National Archives.

'Working Party for the Evaluation of Different Methods of Therapy in Carcinoma of the Bronchus, Memorandum [Draft]', n.d. FD 7/327. UK National Archives.

'Working Party for the Evaluation of Different Methods of Therapy in Carcinoma of the Bronchus, Minutes of the Meeting Held on 23 June 1959', 1959. FD 7/327. UK National Archives.

'Working Party for the Evaluation of Different Methods of Therapy in Carcinoma of the Bronchus, Minutes of the Meeting Held on 24 June 1958', 1958. FD 7/327. UK National Archives.

'Working Party for the Evaluation of Different Methods of Therapy in Carcinoma of the Bronchus, Notes for Discussion', 10 June 1959. FD 7/327. UK National Archives.

Interviews

Mr Roger Abbey Smith, thoracic surgeon, Somerton, Somerset. Interviewer: Carsten Timmermann, 12 December 2005.

Mr John Dark, thoracic surgeon, Wythenshawe Hospital, Manchester. Interviewer: John V. Pickstone, 5 October 1981.
Professor Ray Donnelly, thoracic surgeon, Roy Castle Foundation, Liverpool. Interviewer: Carsten Timmermann, 11 April 2006.
Professor Peter Goldstraw, thoracic surgeon, Brompton Hospital, London. Interviewer: Carsten Timmermann, 28 July 2009.
Professor Nick Thatcher, medical oncologist, Christie Hospital, Manchester. Interviewer: Carsten Timmermann, 12 February 2002.
Professor Tom Treasure, thoracic surgeon, Guy's Hospital, London. Interviewer: Carsten Timmermann, 18 April 2006.

Published Sources

'A King-size Boom in Filter Tips'. *Daily Mirror*, 29 August 1962.
'A National Thoracic Surgical Service (editorial)'. *British Medical Journal* ii (1944): 633–634.
Abbey Smith, Roger. 'Development of Lung Surgery in the United Kingdom'. *Thorax* 37 (1982): 161–168.
———. 'Treatment of Bronchial Carcinoma (letter)'. *Lancet* 288 (1966): 1134–1135.
Abel, Emily and Saskia Subramanian. *After the Cure: The Untold Stories of Breast Cancer Survivors*. New York: NYU Press, 2008.
Aberle, Denise R., Amanda M. Adams, Christine D. Berg, William C. Black, Jonathan D. Clapp, Richard M. Fagerstrom, Ilana F. Gareen, Constantine Gatsonis, Pamela M. Marcus and JoRean D. Sicks. 'Reduced Lung-cancer Mortality with Low-dose Computed Tomographic Screening'. *The New England Journal of Medicine* 365 (2011): 395–409.
Ackerknecht, Erwin H. 'Historical Notes on Cancer'. *Medical History* 2 (1958): 114–119.
———. *Medicine at the Paris Hospital, 1794–1848*. Baltimore: Johns Hopkins Press, 1967.
Adler, Isaac. *Primary Malignant Growths of the Lungs and Bronchi*. New York: Longmans, Green, and Co., 1912.
———. 'The Diagnosis of Malignant Tumors of the Lung'. *New York Medical Journal* 63 (1896): 173–176, 201–209.
Advisory Committee on Breast Cancer Screening. *Screening for Breast Cancer in England: Past and Future*. London: NHS Cancer Screening Programmes, 2006.
Advisory Committee to the Surgeon General of the Public Health Service. *Smoking and Health*. Washington, DC: US Department of Health, Education, and Welfare – Public Health Service, 1964.
Agnew, R. A. L. *The Life of Sir John Forbes (1787–1861)*. Bramber, West Sussex: Bernard Durnford, 2002.
Aisner, Joseph. 'CT Screening for Lung Cancer: Are We Ready for Wide-scale Application?' *Clinical Cancer Research* 13 (2007): 4951–4953.
Allison, P. R. and Leslie J. Temple. 'The Future of Thoracic Surgery'. *Thorax* 21 (1966): 99–103.
'Another Winter of Bronchitis (editorial)'. *British Medical Journal* ii (1959): 1076–1077.
Aronowitz, Robert A. 'The Converged Experience of Risk and Disease'. *The Milbank Quarterly* 87 (2009): 417–442.

Aronowitz, Robert A. 'Do Not Delay: Breast Cancer and Time, 1900–1970'. *Milbank Quarterly* 79 (2001): 355–386.

———. *Unnatural History: Breast Cancer and American Society.* Cambridge & New York: Cambridge University Press, 2007.

'Attitudes to Smoking (editorial)'. *British Medical Journal* ii (1967): 460.

Austoker, Joan. *A History of the Imperial Cancer Research Fund, 1902–1986.* Oxford: Oxford University Press, 1988.

Bahnson, Claus Bahne. 'In Memory of Dr David M. Kissen: His Work and His Thinking'. *Annals of the New York Academy of Sciences* 164 (1969): 313–318.

Baines, Joanna. 'Three Stories: Generations of Breast Cancer'. In *Cancer Patients, Cancer Pathways: Historical and Sociological Perspectives*, edited by Carsten Timmermann and Elizabeth Toon, 13–35. Basingstoke: Palgrave Macmillan, 2012.

Baines, Joanna E. 'Cancer and the Individual in Britain 1850 to 2000'. PhD dissertation, University of Manchester, 2009.

'"Ban the Smokers" Bid in Four Cities'. *Daily Mirror*, 21 March 1962.

Barach, Alvan L. 'Air Pollution and Health: A Review'. *Bulletin of the New York Academy of Medicine* 35 (1959): 493–510.

Barker, D. 'Leave 'em Laughing: Obituary of Roy Castle'. *The Guardian*, 3 September 1994.

Barnes, Emm. 'Between Remission and Cure: Patients, Practitioners and the Transformation of Leukaemia in the Late Twentieth Century'. *Chronic Illness* 3 (2007): 253–264.

———. 'Captain Chemo and Mr Wiggly: Patient Information for Children with Cancer in the Late Twentieth Century'. *Social History of Medicine* 19 (2006): 501–519.

———. 'Caring and Curing: Paediatric Cancer Services Since 1960'. *European Journal of Cancer Care* 14 (2005): 373–380.

Barrett, N. R. 'Introduction'. In *Some Aspects of Carcinoma of the Bronchus and Other Malignant Diseases of the Lung. A Symposium Held at King Edward VII Hospital, Midhurst, July 4th and 5th, 1966*, edited by Douglas Teare and Joan Fenning, 1–4. Midhurst: King Edward VII Hospital, 1966.

Bates, Michael, ed. *Bronchial Carcinoma: An Integrated Approach to Diagnosis and Management.* Berlin: Springer, 1984.

Bayer, Ronald and Jennifer Stuber. 'Tobacco Control, Stigma, and Public Health: Rethinking the Relations'. *American Journal of Public Health* 96 (2006): 47–50.

Bedford, R. 'The Case Against Cigarettes'. *Daily Mirror*, 3 March 1962.

Belcher, J. R. 'Indications for Surgery and Choice of Operation'. In *Some Aspects of Carcinoma of the Bronchus and Other Malignant Diseases of the Lung. A Symposium Held at King Edward VII Hospital, Midhurst, July 4th and 5th, 1966*, edited by Douglas Teare and Joan Fenning, 100–104. Midhurst: King Edward VII Hospital, 1966.

———. 'Thirty Years of Surgery for Carcinoma of the Bronchus'. *Thorax* 38 (1983): 428–432.

———. 'Treatment of Bronchial Carcinoma (letter)'. *Lancet* 288 (1966): 1190–1191.

Bentley, F. J. and Z. A. Leitner. 'Mass Radiography. With Special Reference to Screen Photography and Pulmonary Tuberculosis'. *British Medical Journal* i (1940): 879–883.

Berlivet, Luc. '"Association or Causation?" The Debate on the Scientific Status of Risk Factor Epidemiology, 1947–c. 1965'. *Clio Medica* 75 (2005): 39–74.

Berridge, Virginia, ed. *Making Health Policy: Networks in Research and Policy after 1945*. Amsterdam: Rodopi, 2005.

———. *Marketing Health: Smoking and the Discourse of Public Health in Britain, 1945–2000*. Oxford & New York: Oxford University Press, 2007.

———. 'Passive Smoking and Its Prehistory in Britain: Policy Speaks to Science?' *Social Science and Medicine* 49 (1999): 1183–1195.

———. 'Science and Policy: The Case of Postwar British Smoking Policy'. In *Ashes to Ashes: The History of Smoking and Health*, edited by S. A. Lock, L. A. Reynolds and E. M. Tansey, 143–170. Amsterdam & Atlanta: Rodopi, 1998.

Berridge, Virginia and Kelly Loughlin. 'Smoking and the New Health Education in Britain, 1950s–1970s'. *American Journal of Public Health* 95 (2005): 956–964.

Berridge, Virginia and Suzanne Taylor, eds. *The Big Smoke: Fifty Years after the 1952 London Smog*. Witness Seminar Held 10 December 2002 at the Brunei Gallery, SOAS, London. London: Centre for History in Public Health, London School of Hygiene & Tropical Medicine, 2005. http://history.lshtm.ac.uk/BigSmokeNS.pdf.

Bignall, John. 'John Reginald Bignall (obituary)'. *British Medical Journal* 322 (2001): 176.

Bignall, John R., ed. *Carcinoma of the Lung*. Neoplastic Disease at Various Sites 1. Edinburgh & London: Livingstone, 1958.

———. 'Early Diagnosis of Bronchial Carcinoma'. *British Medical Journal* i (1966): 341–343.

Bignall, John R., Marjorie Martin and David W. Smithers. 'Survival in 6086 Cases of Bronchial Carcinoma'. *Lancet* 289 (1967): 1067–1070.

Black, William C. and John A. Baron. 'CT Screening for Lung Cancer: Spiraling into Confusion? [editorial]'. *JAMA: Journal of the American Medical Association* 297 (2007): 995–997.

Blanshard, Gerald. 'The Palliation of Bronchial Carcinoma by Radiotherapy'. *Lancet* 266 (1955): 897–901.

Bleehen, Norman M. 'Chemotherapy for Carcinoma of the Bronchus – A Brief Review'. *Postgraduate Medical Journal* 49 (1973): 723–728.

Bleehen, S. S. 'Obituary: Norman Montague Bleehen'. *British Medical Journal* 336 (2008): 1383.

Bloor, David. *Knowledge and Social Imagery*. 2nd ed. Chicago: University of Chicago Press, 1991.

Bonser, Georgiana. 'The Incidence of Tumours of the Respiratory Tract in Leeds'. *Journal of Hygiene* 28 (1929): 340–354.

Booth, Christopher C. 'From Art to Science: The Story of Clinical Research'. In *A Physician Reflects: Herman Boerhaave and Other Essays*, 79–101. London: Wellcome Trust Centre for the History of Medicine at UCL, 2003.

———. 'Smoking and the Gold-headed Cane: The Royal College of Physicians Enters the Modern World'. In *A Physician Reflects: Herman Boerhaave and Other Essays*, 155–160. London: Wellcome Trust Centre for the History of Medicine at UCL, 2003.

———. 'Smoking and the Royal College of Physicians'. In *Ashes to Ashes: The History of Smoking and Health*, edited by S. A. Lock, L. A. Reynolds and E. M. Tansey, 192–197. Amsterdam: Rodopi, 1998.

Bouluri, Yvonne. 'Dundee Mum Tells of Lung Cancer Ordeal: I Smoked for 45 Years and Felt Fit as a Fiddle. Until Docs Found My Lung Tumour. More Women in Scotland Are Dying from Lung Cancer Than Ever Before'. *The Scottish Sun*, 10 December 2009. http://www.thescottishsun.co.uk/scotsol/homepage/news/2765369/Dundee-mum-tells-of-lung-cancer-ordeal.html.

Bournemouth University Press Office. 'Kids Worse Off with Smoking Parents'. *Bournemouth University: News and Events*, 1 September 2006. http://www.bournemouth.ac.uk/newsandevents/News/2006/smoking_parents.htm.

Bradford, Sarah. *King George VI*. London: Weidenfeld and Nicolson, 1989.

Brandt, Allan M. *The Cigarette Century: The Rise, Fall, and Deadly Persistence of the Product That Defined America*. New York: Basic Books, 2007.

Brandt, Allan M. and Paul Rozin. *Morality and Health*. New York: Routledge, 1997.

Breslow, Lester and John Goldsmith. 'Health Effects of Air Pollution'. *American Journal of Public Health* 48 (1958): 913–917.

Brett, G. Z. 'Earlier Diagnosis and Survival in Lung Cancer'. *British Medical Journal* iv (1969): 260–262.

Brewer, Lyman A. 'Historical Notes on Lung Cancer before and after Graham's Successful Pneumonectomy in 1933'. *American Journal of Surgery* 143 (1982): 650–659.

Brock, Russell. 'Thoracic Surgery: And the Long-term Results of Operation for Bronchial Carcinoma. The Tudor Edwards Memorial Lecture Delivered at the Royal College of Surgeons of England on 20 May 1964'. *Annals of the Royal College of Surgeons* 35 (1964): 195–213.

Brodsky, Jay B. 'The Evolution of Thoracic Anesthesia'. *Thoracic Surgery Clinic* 15 (2005): 1–10.

Brooks, William D. W., Maurice Davidson, Clement Price Thomas, Kenneth Robson and David W. Smithers. 'Carcinoma of the Bronchus: A Report on the First Five Years' Work of the Joint Consultation Clinic for Neoplastic Diseases of the Brompton Hospital and the Royal Cancer Hospital'. *Thorax* 6 (1951): 1–16.

Brunning, D. A. and C. E. Dukes. 'The Origin and Early History of the Institute of Cancer Research of the Royal Cancer Hospital'. *Proceedings of the Royal Society of Medicine* 58 (1965): 33–36.

Bryder, Linda. *Below the Magic Mountain: A Social History of Tuberculosis in Twentieth-century Britain*. Oxford & New York: Clarendon Press, 1988.

Buchanan, Roderick D. *Playing with Fire: The Controversial Career of Hans J. Eysenck*. Oxford: Oxford University Press, 2010.

'Budget Accounts: On Schedule'. *Economist* (6 October 1962): 74–75.

Burnet, N. G., S. J. Jefferies, R. J. Benson, D. P. Hunt and F. P. Treasure. 'Years of Life Lost (YLL) from Cancer Is an Important Measure of Population Burden – and Should Be Considered When Allocating Research Funds'. *British Journal of Cancer* 92 (2005): 241–245.

Burnham, John C. 'American Physicians and Tobacco Use: Two Surgeons General, 1929 and 1964'. *Bulletin of the History of Medicine* 63 (1989): 1–31.

Burnham, John C. *Bad Habits: Drinking, Smoking, Taking Drugs, Gambling, Sexual Misbehavior, and Swearing in American History*. New York: New York University Press, 1993.

C. D., J. G. S. and A. T. J. 'Sir Clement Price Thomas (obituary)'. *British Medical Journal* i (1973): 807–808.

'Cancer of the Lung (editorial)'. *Lancet* 210 (1927): 125.

Cancer Research UK. 'CancerStats Key Facts'. Accessed 14 August 2012. http://info.cancerresearchuk.org/cancerstats/keyfacts/.

———. 'Liver Cancer Statistics – UK'. Accessed 18 July 2012. http://info.cancerresearchuk.org/cancerstats/types/liver/.

———. 'Lung Cancer and Smoking'. Accessed 18 July 2012. http://info.cancerresearchuk.org/cancerstats/types/lung/.

———. 'Pancreatic Cancer Statistics'. Accessed 18 July 2012. http://info.cancerresearchuk.org/cancerstats/types/pancreas/.

Cantor, David. 'Cancer'. In *Companion Encyclopedia of the History of Medicine*, edited by W. F. Bynum and Roy Porter, 1: 537–561. London: Routledge, 1993.

———. , ed. *Cancer in the Twentieth Century*. Baltimore: Johns Hopkins University Press, 2008.

———. 'The Definition of Radiobiology: The Medical Research Council's Support for Research into the Biological Effects of Radiation in Britain, 1919–1939'. PhD dissertation, University of Lancaster, 1987.

Carmichael, J. 'Introduction'. *Annals of Oncology* 6, Supplement 1 (1995): S1–S2.

Carr, David T. 'A Report on the Development of the Staging System for Cancer of the Lung'. In *Sixth National Cancer Conference Proceedings*, 877–878. Philadelphia: J. B. Lippincott, 1970.

———. 'Classification of Lung Cancer'. In *Lung Cancer: Current Status and Prospects for the Future. The University of Texas M.D. Anderson Hospital and Tumor Institute at Houston Twenty-eighth Annual Clinical Conference on Cancer*, edited by Clifton F. Mountain and David T. Carr, 39–48. Austin: University of Texas Press, 1986.

Cartwright, Ann and F. M. Martin. 'Some Popular Beliefs Concerning the Causes of Cancer'. *British Medical Journal* ii (1958): 592–594.

Cartwright, Ann, F. M. Martin and J. G. Thomson. 'Distribution and Development of Smoking Habits'. *Lancet* 274 (1959): 725–727.

———. 'Efficacy of an Anti-Smoking Campaign'. *Lancet* 275 (1960): 327–330.

———. 'Health Hazards of Cigarette Smoking: Current Popular Beliefs'. *British Journal of Preventive and Social Medicine* 14 (1960): 160–166.

Castle, Roy. *Now and Then: An Autobiography*. London: Macmillan, 1994.

Centers for Disease Control and Prevention. 'CDC – Lung Cancer Statistics'. Accessed 18 July 2012. http://www.cdc.gov/cancer/lung/statistics/.

Chalmers, I. and M. Clarke. 'J Guy Scadding and the Move from Alternation to Randomization'. *James Lind Library*, 2002. http://www.jameslindlibrary.org/illustrating/articles/j-guy-scadding-and-the-move-from-alternation-to-ran-domization.

Chapple, A., S. Ziebland and A. McPherson. 'Stigma, Shame, and Blame Experienced by Patients with Lung Cancer: Qualitative Study'. *British Medical Journal* 328 (2004): 1470–1473.

Charlton, John and Mike Murphy, eds. *The Health of Adult Britain 1841–1994*. London: The Stationery Office, 1997.

Charmaz, Kathy. *Good Days, Bad Days: The Self in Chronic Illness and Time*. New Brunswick, N.J: Rutgers University Press, 1991.

Charteris and Owen Williams. 'Glasgow Royal Infirmary: Cancer of Lung and Pleuro-pneumonia'. *Lancet* 112 (1878): 732–733.

'Chest Surgery Film'. *British Medical Journal* ii (1943): 397.

'Cigarette Slot Machine Ban'. *Daily Mirror*, 6 April 1962.

'Cigarette TV Ban – and It's Just the First Step'. *Daily Mirror*, 9 February 1965.

'Cigarettes and Cancer (editorial)'. *British Medical Journal* ii (1950): 767–768.

'Cigarettes in School Tuck Boxes'. *Daily Mirror*, 29 December 1960.

'Cigarettes on the Counter'. *Economist* (15 January 1955): 211–212.

'Cigarettes: No Drastic Steps Likely'. *Daily Mirror*, 9 March 1962.

Clark, D. 'Cradled to the Grave? Terminal Care in the United Kingdom, 1948–67'. *Mortality* 4 (1999): 225–247.

Clark, David. 'From Margins to Centre: A Review of the History of Palliative Care in Cancer'. *Lancet Oncology* 8 (2007): 430–438.

Clark, David, Neil Small, Michael Wright, Michelle Winslow and Nic Hughes. *A Bit of Heaven for the Few?: An Oral History of the Modern Hospice Movement in the United Kingdom*. Observatory Publications, 2005.

Clemmesen, Johannes. 'Lung Cancer from Smoking: Delays and Attitudes, 1912–1965'. *American Journal of Industrial Medicine* 23 (1993): 941–953.

Cole, Harvey. 'The Economic Effects'. In *Common Sense About Smoking*, edited by Charles M. Fletcher, Harvey Cole, Lena Jeger and Christopher Wood, 43–75. Harmondsworth: Penguin, 1963.

Collins, Harry M. and Trevor J. Pinch. *The Golem*. Cambridge: Cambridge University Press, 1994.

Commission on Chronic Illness. *Prevention of Chronic Illness*. Vol. 2. Chronic Illness in the United States. Cambridge: Harvard University Press, 1957.

Conrad, Karen M., Brian R. Flay and David Hill. 'Why Children Start Smoking Cigarettes: Predictors of Onset'. *British Journal of Addiction* 87 (1992): 1711–1724.

'Consumption is Rising'. *Economist* (14 September 1957): 874.

Cook, J. W. 'Ernest Laurence Kennaway. 1881–1958'. *Biographical Memoirs of Fellows of the Royal Society* 4 (1958): 139–154.

Cooter, Roger. 'Keywords in the History of Medicine: Teamwork'. *The Lancet* 363 (2004): 1245.

Copeland, Murray M. 'American Joint Committee on Cancer Staging and End Results Reporting: Objectives and Progress'. *Cancer* 18 (1965): 1637–1640.

Cox, Malcolm and Colin Pritchard. 'Smoking Parents and the Drink, Drug and Sexual Activity of Their Children'. In *Adolescence and Alcohol: An International Perspective*, edited by Isack Kandel, Joav Merrick and Leo Sher, 125–137. London: Freund, 2006.

Cox, Robert. 'In Memoriam: Sir Stanford Cade, KBE CB'. *Annals of the Royal College of Surgeons of England* 54 (1974): 94–96.

Crafoord, Clarence. *On the Technique of Pneumonectomy in Man: A Critical Survey of the Experimental and Clinical Development and a Report of the Author's Material and Technique*. Tryckeri Aktiebolaget Thule, 1938.

Craig, Mary. *The Last Freedom*. London: Hodder & Stoughton, 1997.

Crofton, John. 'John Guyett Scadding (obituary)'. *British Medical Journal* 320 (2000): 189.

Cruickshanks, K. J., R. Klein, B. E. Klein, T. L. Wiley, D. M. Nondahl and T. S. Tweed. 'Cigarette Smoking and Hearing Loss: The Epidemiology of Hearing Loss Study'. *JAMA: The Journal of the American Medical Association* 279 (1998): 1715–1719.

Cunningham, George J. *The History of British Pathology*. Bristol: White Tree Books, 1992.

Curran, W. 'A Puzzling Case of Cancer of the Lung'. *Lancet* 116 (1880): 258–259.

Cuthbertson, David. 'Historical Notes on the Origin of the Association Between Lung Cancer and Smoking'. *Journal of the Royal College of Physicians of London* 2 (1967): 191–196.

Daley, Allen. 'The Health of the Nation'. *British Medical Journal* i (1951): 1279–1285.

Davidson, Maurice and F. G. Rouvray. *The Brompton Hospital: The Story of a Great Adventure*. London: Lloyd-Luke, 1954.

Davies, Hugh Morriston. 'Recent Advances in the Surgery of the Lung and Pleura'. *British Journal of Surgery* 1 (1913): 228–258.

———. 'The Mechanical Control of Pneumothorax During Operations on the Chest, with a Description of a New Apparatus'. *British Medical Journal* ii (1911): 61–66.

Davies, Hugh Morriston and Robert Coope, eds. *War Injuries of the Chest*. Edinburgh, 1942.

Davies, Robert Price. *Baguley and Wythenshawe Hospitals: A History*. Manchester: R. P. Davies, 2002.

———. 'Memories Provided by John Dark – Chest and Heart Surgeon, Baguley and Wythenshawe Hospitals'. In *Baguley and Wythenshaw Hospitals: A History*, 136–148. Manchester: R. P. Davies, 2002.

Deeley, T. J. 'The Early Diagnosis of Lung Cancer'. *Postgraduate Medical Journal* 48 (1972): 33–45.

Department of Health. *Statistics on Smoking: England, 1978 Onwards*. Statistical Bulletin, 20 July 2000. http://www.dh.gov.uk/en/Publicationsandstatistics/Statistics/StatisticalWorkAreas/Statisticalpublichealth/DH_4015501.

'Dimbleby Cancer Care'. Homepage. Accessed 14 August 2012. http://www.dimblebycancercare.org/.

Dingle, John H. 'Studies of Respiratory and Other Illnesses in Cleveland (Ohio) Families'. *Proceedings of the Royal Society of Medicine* 49 (1956): 5–7.

Doll, Richard. 'Austin Bradford Hill, 8 July 1897–18 April 1991'. *Biographical Memoirs of Fellows of the Royal Society* 40 (1994): 129–140.

———. 'Commentary: Lung Cancer and Tobacco Consumption'. *International Journal of Epidemiology* 30 (2001): 30–31.

———. 'In Memoriam: Ernst Wynder, 1923–1999'. *American Journal of Public Health* 89 (1999): 1798–1799.

———. 'Occupational Lung Cancer: A Review'. *British Journal of Industrial Medicine* 16 (1959): 181–190.

———. 'Stocks, Percy (1889–1974)'. In *Oxford Dictionary of National Biography*. Oxford: Oxford University Press, 2004. http://www.oxforddnb.com/view/article/57335.

———. 'The First Reports on Smoking and Lung Cancer'. In *Ashes to Ashes: The History of Smoking and Health*, edited by S. A. Lock, L. A. Reynolds and E. M. Tansey, 130–140. Amsterdam & Atlanta: Rodopi, 1998.

———. 'The Pierre Denoix Memorial Lecture: Nature and Nurture in the Control of Cancer'. *European Journal of Cancer* 35 (1999): 16–23.

———. 'Uncovering the Effects of Smoking: Historical Perspective'. *Statistical Methods in Medical Research* 7 (1998): 87–117.

Doll, Richard and Austin Bradford Hill. 'Smoking and Carcinoma of the Lung'. *British Medical Journal* ii (1950): 739–748.

————. 'The Mortality of Doctors in Relation to Their Smoking Habits: A Preliminary Report'. *British Medical Journal* i (1954): 1451–1455.

Doll, Richard, Richard Peto, J. Boreham and I. Sutherland. 'Mortality from Cancer in Relation to Smoking: 50 Years Observations on British Doctors'. *British Journal of Cancer* 92 (2005): 426–439.

Donnelly, Ray. *Cinderella Cancer: A Personal History of the Roy Castle Lung Cancer Foundation*. Liverpool: Bluecoat Press, 2006.

Drayson, D. Droytt. 'Non-Smoking Carriages (Letter to the Editor)'. *The Times*, 11 August 1942.

'Drinks and Smokes'. *Economist* (17 November 1956): 585.

Duffin, Jacalyn. *To See with a Better Eye: A Life of R.T.H. Laennec*. Princeton, NJ: Princeton University Press, 1998.

Duguid, J. B. 'The Incidence of Intrathoracic Tumours in Manchester'. *Lancet* 210 (1927): 111–116.

Duke, Simon L. and Tim Eisen. 'Finding Needles in a Haystack: Annual Low-dose Computed Tomography Screening Reduces Lung Cancer Mortality in a High-risk Group'. *Expert Review of Anticancer Therapy* 11 (2011): 1833–1836.

'Early Diagnosis of Lung Cancer (editorial)'. *British Medical Journal* i (1968): 710–711.

Edey, A. J. and D. M. Hansell. 'CT Lung Cancer Screening in the UK'. *British Journal of Radiology* 82 (2009): 529–531.

Ehrenreich, Barbara. 'Smile! You've Got Cancer'. *The Guardian*, 2 January 2010.

Elliot, Rosemary. 'Growing up and Giving Up: Smoking in Paul Thompson's 100 Families'. *Oral History* 29 (2001): 73–84.

Ellis, J. R. C. and F. V. Gleeson. 'Lung Cancer Screening'. *British Journal of Radiology* 74 (2001): 478–485.

Ellman, Philip. *Essentials in Diseases of the Chest for Students and Practitioners*. Oxford & New York: Oxford University Press, 1952.

Epstein, Steven. *Impure Science: AIDS, Activism, and the Politics of Knowledge*. Berkeley: University of California Press, 1996.

'Experimental Links Between Tobacco and Lung Cancer (editorial)'. *British Medical Journal* i (1958): 1050–1051.

Eyler, John M. *Sir Arthur Newsholme and State Medicine, 1885–1935*. Cambridge: Cambridge University Press, 1997.

————. *Victorian Social Medicine: The Ideas and Methods of William Farr*. Baltimore & London: Johns Hopkins University Press, 1979.

Eysenck, H. J. *Smoking, Health and Personality*. London: Four Square, 1966.

Eysenck, H. J., Mollie Tarrant and Myra Woolf. 'Smoking and Personality'. *British Medical Journal* i (1960): 1456–1460.

Field, J. K., D. L. Smith, S. Duffy and A. Cassidy. 'The Liverpool Lung Project Research Protocol'. *International Journal of Oncology* 27 (2005): 1633–1645.

Field, John K. 'Lung Cancer Risk Models Come of Age'. *Cancer Prevention Research* 1 (2008): 226–228.

Field, John K. and S. W. Duffy. 'Lung Cancer Screening: The Way Forward'. *British Journal of Cancer* 99 (2008): 557–562.

Field, John K. and J. H. Youngson. 'The Liverpool Lung Project: A Molecular Epidemiological Study of Early Lung Cancer Detection'. *European Respiratory Journal* 20 (2002): 464–479.

Fisher, R. A. *Smoking: The Cancer Controversy*. Edinburgh: Oliver & Boyd, 1958.

Fletcher, Charles M. 'Chronic Disabling Respiratory Disease: Ends and Means of Study'. *California Medicine* 88 (1958): 1–11.

Fletcher, Charles M., P. C. Elmes, A. S. Fairbairn and C. H. Wood. 'The Significance of Respiratory Symptoms and the Diagnosis of Chronic Bronchitis in a Working Population'. *British Medical Journal* ii (1959): 257–266.

Fontana, Robert S., David R. Sanderson, Lewis B. Woolner, William F. Taylor, W. Eugene Miller, John R. Muhm, Philip E. Bernatz, W. Spencer Payne, Peter C. Pairolero and Erik J. Bergstralh. 'Screening for Lung Cancer: A Critique of the Mayo Lung Project'. *Cancer* 67 (1991): 1155–1164.

Forbes, John. *Original Cases with Dissections and Observations Illustrating the Use of the Stethoscope and Percussion in the Diagnosis of Diseases of the Chest*. London: Underwood, 1824.

Foster, William Derek. *A Short History of Clinical Pathology*. Edinburgh & London: E. & S. Livingstone, 1961.

Foucault, Michel. *The Birth of the Clinic: An Archaeology of Medical Perception*. New York: Vintage Books, 1975.

Fox, M. S. 'On the Diagnosis and Treatment of Breast Cancer'. *JAMA: The Journal of the American Medical Association* 241 (1979): 489–494.

Fox, Wallace and J. G. Scadding. 'Medical Research Council Comparative Trial of Surgery and Radiotherapy for Primary Treatment of Small-Celled or Oat-Celled Carcinoma of Bronchus: Ten-year Follow-up'. *Lancet* 302 (1973): 63–65.

Frank, Arthur W. *At the Will of the Body: Reflections on Illness*. Boston: Houghton Mifflin, 2002.

———. *The Wounded Storyteller: Body, Illness, and Ethics*. Chicago: University of Chicago Press, 1995.

Fry, John. 'Chronic Bronchitis in General Practice'. *British Medical Journal* i (1954): 190–194.

Gardner, Kirsten E. *Early Detection: Women, Cancer, & Awareness Campaigns in the Twentieth-century United States*. Chapel Hill: University of North Carolina Press, 2006.

Garfield, Simon. 'The Man Who Saved a Million Lives'. *Observer Magazine* (24 April 2005): 33–39.

Garfinkel, Lawrence. 'Classics in Oncology: E. Cuyler Hammond, ScD (1912–1986)'. *CA: A Cancer Journal for Clinicians* 38 (1988): 23–27.

Gaudillière, Jean-Paul. 'Cancer'. In *The Modern Biological and Earth Sciences*, edited by Peter Bowler and John Pickstone, 6: 486–503. The Cambridge History of Science. Cambridge: Cambridge University Press, 2009.

———. 'The Molecularization of Cancer Etiology in the Postwar United States: Instruments, Politics and Management'. In *Molecularizing Biology and Medicine: New Practices and Alliances, 1910s–1970s*, edited by Soraya de Chadarevian and Harmke Kamminga, 461–477. Amsterdam: Harwood Academic, 1998.

Glantz, Leonard. 'Should Smokers Be Refused Surgery? No'. *British Medical Journal* 334 (2007): 21.

Glantz, Stanton A. and Edith D. Balbach. *Tobacco War: Inside the California Battles*. Berkeley: University of California Press, 2000.

Goddard, Jacqui. 'The Winston Man Dies of Lung Cancer ... Just One Month before He Was Due to Testify Against Big Tobacco Company'. *Mail Online*, 3 March 2009. http://www.dailymail.co.uk/news/article-1158932/The-

Winston-Man-dies-lung-cancer–just-month-testify-big-tobacco-company.html.

Goffman, Erving. *Stigma: Notes on the Management of Spoiled Identity.* Harmondsworth: Penguin, 1968.

Golding, Stephen J. 'The Role of Computer Tomography in the Management of Bronchial Carcinoma'. In *Bronchial Carcinoma: An Integrated Approach to Diagnosis and Management*, edited by Michael Bates, 97–113. Berlin: Springer, 1984.

Goldstraw, Peter and John J. Crowley. 'The International Association for the Study of Lung Cancer International Staging Project on Lung Cancer'. *Journal of Thoracic Oncology* 1 (2006): 281–286.

Graham, Evarts A. 'Changing Concepts in Surgery'. *Postgraduate Medicine* 7 (1950): 154–156.

Graham, Evarts A. and J. J. Singer. 'Successful Removal of an Entire Lung for Carcinoma of the Bronchus'. *Journal of the American Medical Association* 101 (1933): 1371–1374.

Gray, Brenda. 'Sputum Cytodiagnosis in Bronchial Carcinoma'. *Lancet* 284 (1964): 549–552.

Gruhn, John G. 'A History of the Histopathology of Lung Cancer'. In *Lung Cancer: The Evolution of Concepts*, edited by John G. Gruhn and Steven T. Rosen, 1: 25–63. New York: Field & Wood Medical Publishers, 1989.

Halnan, K. E. 'Fifty Years of the National Health Service 1948–1998: A Personal History of Progress in the Treatment of Cancer'. *Clinical Oncology* 11 (1999): 55–60.

———. 'Obituary: Dr Constance A. P. Wood'. *BIR Bulletin* (March 1986): B16–B17.

Hammond, E. Cuyler and Daniel Horn. 'The Relationship Between Human Smoking Habits and Death Rates: A Follow-up Study of 187,766 Men'. *Journal of the American Medical Association* 155 (1954): 1316–1328.

Hansen, Heine H. *History of the IASLC 1972–2007.* Orange Park, FL: Editorial Rx Press for the IASLC, 2009.

———. 'International Association for the Study of Lung Cancer: The First Three Decades'. *Journal of Thoracic Oncology* 1 (2006): 3–4.

Harkness, Jon M. 'The U.S. Public Health Service and Smoking in the 1950s: The Tale of Two More Statements'. *Journal of the History of Medicine and Allied Sciences* 62 (2006): 171–212.

Harnett, W. L. *A Survey of Cancer in London: Report of the Clinical Cancer Research Committee.* London: British Empire Cancer Campaign, 1952.

Harris, Gardiner. 'Cigarette Company Paid for Lung Cancer Study'. *New York Times*, 26 March 2008.

Harrison, Kent. 'Treatment of Bronchial Carcinoma (letter)'. *Lancet* 288 (1966): 1254.

Hebblethwaite, Peter. 'Obituary: The Most Rev Derek Worlock'. *The Independent*, 9 February 1996.

Henk, J. M. 'Obituary: Professor Sir David Smithers'. *The Independent*, 5 August 1995.

'Henry Marshall Hughes'. *Lives of the Fellows of the Royal College of Physicians (Munk's Roll)* 4 (1955): 36–37.

Henschke, Claudia I. and Peggy McCarthy. *Lung Cancer: Myths, Facts, Choices – and Hope.* New York & London: Norton, 2002.

Henschke, Claudia I., Dorothy I. McCauley, David F. Yankelevitz, David P. Naidich, Georgeann McGuiness, Olli S. Miettinen, Daniel M. Libby, Mark W. Pasmantier, June Koizumi, Nasser K. Altorki and James P. Smith. 'Early Lung Cancer Action Project: Overall Design and Findings from Baseline Screening'. *Lancet* 354 (1999): 99–105.

Henschke, Claudia I., David F. Yankelevitz, Daniel M. Libby, Mark W. Pasmantier, James P. Smith and Olli S. Miettinen. 'Survival of Patients with Stage I Lung Cancer Detected on CT Screening'. *New England Journal of Medicine* 355 (2006): 1763–1771.

Herbert, Ian. 'Head of Cancer Fund Quits over Donations'. *The Independent*, 28 October 1999.

Herek, Gregory M. 'AIDS and Stigma'. *American Behavioral Scientist* 42 (1999): 1106–1116.

Higgins, Ian T. T. 'Respiratory Symptoms, Bronchitis, and Ventilatory Capacity in Random Sample of an Agricultural Population'. *British Medical Journal* ii (1957): 1198–1203.

———. 'Tobacco Smoking, Respiratory Symptoms, and Ventilatory Capacity: Studies in Random Samples of the Population'. *British Medical Journal* i (1959): 325–329.

Higgins, Ian T. T., P. D. Oldham, A. L. Cochrane and J. C. Gilson. 'Respiratory Symptoms and Pulmonary Disability in an Industrial Town: Survey of a Random Sample of the Population'. *British Medical Journal* ii (1956): 904–910.

Hill, Austin Bradford. 'Obituary: Dr Percy Stocks, 1889–1974'. *Journal of the Royal Statistical Society. Series A (General)* 138 (1975): 273–274.

Hill, Gerry, Wayne Millar and James Connelly. '"The Great Debate": Smoking, Lung Cancer, and Cancer Epidemiology'. *Canadian Bulletin of Medical History* 20 (2003): 367–386.

Hilton, Matthew. *Smoking in British Popular Culture 1800–2000*. Manchester: Manchester University Press, 2000.

Hirayama, T. 'Non-smoking Wives of Heavy Smokers Have a Higher Risk of Lung Cancer: A Study from Japan'. *British Medical Journal* 282 (1981): 183–185.

Hoffmann, Dietrich and Ilse Hoffmann. 'Obituary: Ernst L. Wynder MD Dr Sc Hc (mult) Dr Med Hc, 1922–1999'. *Tobacco Control* 8 (1999): 444–445.

Holmes Sellors, Sir Thomas. 'The Management of Cancer of the Lung'. *The Practitioner* 196 (1966): 17–22.

Honess, Davina. 'Obituary: Norman Bleehen'. *British Journal of Cancer* 99 (2008): 681–682.

Hooper, Paul. *Smoking Issues: A Quick Guide*. Cambridge: Daniels Publishing, 1995.

Howell, J. D. '"Soldier's Heart": The Redefinition of Heart Disease and Specialty Formation in Early Twentieth-century Great Britain'. *Medical History. Supplement* 5 (1985): 34–52.

Huggins, A., B. B. Muir, P. T. Donnan, W. Hepburn, R. J. Prescott, T. A. Anderson, J. Lamb, F. E. Alexander, U. Chetty, Patrick Forrest and A. E. Kirkpatrick. 'Edinburgh Trial of Screening for Breast Cancer: Mortality at Seven Years'. *The Lancet* 335 (1990): 241–246.

Hughes, Henry Marshall. 'Cases of Malignant Disease of the Lung'. *Guy's Hospital Reports* 6 (1841): 330–346.

Hurt, Raymond. *The History of Cardiothoracic Surgery: From Early Times*. New York & London: Parthenon, 1996.

Ikeda, Shigeto. 'Flexible Bronchofibrescope'. *Annals of Otology, Rhinology and Laryngology* 79 (1970): 916–919.

'Imperial Tobacco'. *Economist* (28 July 1962): 400.

'In Memoriam: Arthur Tudor Edwards (1890–1946)'. *British Journal of Surgery* 34 (1946): 206–207.

'International Association for the Study of Lung Cancer'. *Journal of Thoracic Oncology* 3 Suppl. (2008): S1–S4.

'Investigators Took Their Own Advice'. *The Times*, 8 March 1962.

'Is Consumption Falling Off?' *Economist* (15 March 1958): 959.

Israel, L. 'Chemotherapy in Inoperable Bronchial Carcinoma (letter)'. *Lancet* 297 (1971): 971–972.

Jacyna, L. S. 'The Laboratory and the Clinic: The Impact of Pathology on Surgical Diagnosis in the Glasgow Western Infirmary, 1875–1910'. *Bulletin of the History of Medicine* (1988): 384–406.

Jeger, Lena. 'The Social Implications'. In *Common Sense About Smoking*, edited by Charles M. Fletcher, Harvey Cole, Lena Jeger and Christopher Wood, 76–106. Harmondsworth: Penguin, 1963.

Jemal, Ahmedin, Michael J. Thun, Lynn A. G. Ries, Holly L. Howe, Hannah K. Weir, Melissa M. Center, Elizabeth Ward, Xiao-Cheng Wu, Christie Eheman, Robert Anderson, Umed A. Ajani, Betsy Kohler and Brenda K. Edwards. 'Annual Report to the Nation on the Status of Cancer, 1975–2005, Featuring Trends in Lung Cancer, Tobacco Use, and Tobacco Control'. *Journal of the National Cancer Institute* 100 (2008): 1672–1694.

Jett, James R. 'The New Official Journal of the IASLC'. *Journal of Thoracic Oncology* 1 (2006): 1–2.

Joint National Cancer Survey Committee of the Marie Curie Memorial and the Queen's Institute of District Nursing. *Report on a National Survey Concerning Patients with Cancer Nursed at Home*. London: Marie Curie Memorial, 1952.

Jones, Arthur. 'Megavoltage Radiotherapy of Carcinoma of the Bronchus'. In *Some Aspects of Carcinoma of the Bronchus and Other Malignant Diseases of the Lung. A Symposium Held at King Edward VII Hospital, Midhurst, July 4th and 5th, 1966*, edited by Douglas Teare and Joan Fenning, 95–99. Midhurst: King Edward VII Hospital, 1966.

Joules, Horace. 'Liability to Lung Cancer (Letter to the Editor)'. *The Times*, 9 February 1956.

———. 'Symposium on Cancer of the Lung and Tobacco Consumption'. *British Journal of Addiction* 53(1) (1956): 17–22.

Kabat, Geoffrey C. *Hyping Health Risks: Environmental Hazards in Daily Life and the Science of Epidemiology*. New York: Columbia University Press, 2008.

Karnofsky, David A., Walter H. Abelmann, Lloyd F. Craver and Joseph H. Burchenal. 'The Use of the Nitrogen Mustards in the Palliative Treatment of Carcinoma. With Particular Reference to Bronchogenic Carcinoma'. *Cancer* 1 (1948): 634–656.

Keating, Conrad. *Smoking Kills: The Revolutionary Life of Richard Doll*. Oxford: Signal, 2009.

Keating, Peter and Alberto Cambrosio. *Cancer on Trial: Oncology as a New Style of Practice*. Chicago, Ill.: University of Chicago Press, 2012.

Kennaway, N. M. and E. L. Kennaway. 'A Further Study of the Incidence of Cancer of the Lung and Larynx'. *British Journal of Cancer* 1 (1947): 260–298.

———. 'A Study of the Incidence of Cancer of the Lung and Larynx'. *Journal of Hygiene* 36(2) (1936): 236–267.

King, George and Arthur Newsholme. 'On the Alleged Increase in Cancer'. *Proceedings of the Royal Society of London* 54 (1893): 209–242.

King, Samantha. *Pink Ribbons, Inc.: Breast Cancer and the Politics of Philanthropy.* Minneapolis: University of Minnesota Press, 2006.

Kirby, Bryan and Caroline Richmond. 'Obituary: Keith Ball: Cofounder of Action on Smoking and Health and the Coronary Prevention Group'. *British Medical Journal* 336 (2008): 452.

Kleinman, Arthur. *The Illness Narratives: Suffering, Healing, and the Human Condition.* New York: Basic Books, 1988.

Kluger, Richard. *Ashes to Ashes: America's Hundred-year Cigarette War, the Public Health, and the Unabashed Triumph of Philip Morris.* New York: Alfred A. Knopf, 1996.

Kmietowicz, Zosia. 'Research Spending on Cancer Doesn't Match Their Death Rates'. *British Medical Journal* 325 (2002): 920.

Kolata, Gina. 'Researchers Dispute Benefits of CT Scans for Lung Cancer'. *New York Times,* 7 March 2007.

———. 'Study Says Beetter Scans May Discover Lung Cancer Sooner'. *New York Times,* 26 October 2006.

———. 'Study Sets Off Debate Over Mammograms' Value'. *New York Times,* 9 December 2001.

Kotin, Paul and Hans L. Falk. 'Air Pollution and Its Effect on Health'. *California Medicine* 82 (1955): 19–24.

Krueger, Gretchen M. *Hope and Suffering: Children, Cancer, and the Paradox of Experimental Medicine.* Baltimore: Johns Hopkins University Press, 2008.

Kubik, Antonin. 'Screening for Lung Cancer High-risk Groups (letter)'. *British Medical Journal* ii (1970): 666.

Lachmund, Jens. *Der abgehorchte Körper: Zur historischen Soziologie der medizinischen Untersuchung.* Opladen: Westdeutscher Verlag, 1997.

Laennec, René T. H. *A Treatise on the Diseases of the Chest, in Which They Are Described According to Their Anatomical Characters, and Their Diagnosis, Established on a New Principle by Means of Acoustick Instruments.* London: Underwood, 1821.

———. 'Encéphaloides'. In *Dictionaire Des Sciences Médicales, Par Une Société de Médecins et de Chirurgiens,* 165–178. Paris: C. L. F. Panckoucke, 1815.

Laing, A. H., R. J. Berry, C. R. Newman and P. Smith. 'Treatment of Small-Cell Carcinoma of Bronchus'. *Lancet* 305 (1975): 129–132.

'Landmark for Keyhole Surgery'. *The Times,* 9 April 1991.

Laszlo, John. *The Cure of Childhood Leukemia: Into the Age of Miracles.* New Brunswick, N.J: Rutgers University Press, 1995.

Lawrence, Christopher. 'Democratic, Divine and Heroic: The History and Historiography of Surgery'. In *Medical Theory, Surgical Practice: Essays in the History of Surgery,* edited by Christopher Lawrence, 1–47. London & New York: Routledge, 1992.

———. , ed. *Medical Theory, Surgical Practice: Essays in the History of Surgery.* London & New York: Routledge, 1992.

Lederer, Susan E. *Flesh and Blood: Organ Transplantation and Blood Transfusion in Twentieth-century America*. Oxford & New York: Oxford University Press, 2008.

Lenfant, Claude. 'Shattuck Lecture: Clinical Research to Clinical Practice – Lost in Translation?' *The New England Journal of Medicine* 349 (2003): 868–874.

Lerner, Barron H. *The Breast Cancer Wars: Hope, Fear, and the Pursuit of a Cure in Twentieth-century America*. New York: Oxford University Press, 2001.

———. *When Illness Goes Public: Celebrity Patients and How We Look at Medicine*. Baltimore: Johns Hopkins University Press, 2006.

'Liability to Lung Cancer: Incidence of Disease "Frightening"'. *The Times*, 31 January 1956.

Lickint, Fritz. *Lungenkrebs der Raucher*. Berlin: Verlag Volk und Gesundheit, 1958.

———. *Tabakgenuß und Gesundheit*. Hannover: Bruno Wilkens Verlag, 1936.

———. *Zigarette und Lungenkrebs*. Hamm (Westf.): Hoheneck-Verlag, 1957.

Lilienfeld, David E. and Paul D. Stolley. *Foundations of Epidemiology*. Third. New York & Oxford: Oxford University Press, 1994.

Lilienthal, Howard. 'Resection of the Lung for Suppurative Infections with a Report Based on 31 Operative Cases in Which Resection Was Done or Intended'. *Annals of Surgery* 75 (1922): 257–320.

'Link Between Smoking and Cancer. Government Acts: Tobacco Firms' Research Offer'. *Manchester Guardian*, 13 February 1954.

Lister, Sam. 'NHS Accused over Women's Breast Cancer Screening Risks'. *The Times*, 20 February 2009.

Lock, Stephen, L. A. Reynolds and E. M. Tansey, eds. *Ashes to Ashes: The History of Smoking and Health*. Amsterdam: Rodopi, 1998.

Logan, Andrew. 'The Beginnings of Thoracic Surgery'. *South African Journal of Thoracic Surgery* 24 (1986): 136–138.

Lorenz, Egon. 'Radioactivity and Lung Cancer: A Critical Review of Lung Cancer in the Miners of Schneeberg and Joachimsthal'. *Journal of the National Cancer Institute* 5 (1944): 1–15.

Löwy, Ilana. 'Cancer: The Century of the Transformed Cell'. In *Science in the Twentieth Century*, edited by John Krige and Dominique Pestre, 461–477. Amsterdam: Harwood Academic, 1997.

———. *Preventive Strikes: Women, Precancer, and Prophylactic Surgery*. Baltimore: Johns Hopkins University Press, 2010.

'Lulling the Smoker'. *Economist* (18 May 1963): 692.

'Lung Cancer and Smoking (editorial)'. *British Medical Journal* i (20 February 1954): 445.

'Lung Cancer Increase "Due to Smoking": Medical Findings Accepted by Government. Local Authorities Asked to Make Facts Known'. *The Times*, 28 June 1957.

Mackenzie, G. Hunter. *A Practical Treatise on the Sputum. With Special Reference to the Diagnosis, Prognosis, and Therapeusis of Diseases of the Throat and Lungs*. Edinburgh & London: Johnston, 1886.

Mackenzie, Sholto. *Cancer: An Inquiry into the Extent to Which Patients Receive Treatment*. Reports on Public Health and Medical Subjects. London: Ministry of Health, 1939.

Macmillan Cancer Support. 'The Cancer Survival Lottery'. Press release, 22 November 2011. http://www.macmillan.org.uk/Aboutus/News/Latest_News/TheCancerSurvivalLottery.aspx.

'Maggie's'. Homepage. Accessed 14 August 2012. http://www.maggiescentres.org/.

Magill, Ivan W. 'An Appraisal of Progress in Anaesthetics. Frederick Hewitt Lecture Delivered to the Royal College of Surgeons of England on 17th March 1965'. *Annals of the Royal College of Surgeons* 38 (1966): 154–165.

———. 'Anaesthesia in Thoracic Surgery, with Special Reference to Lobectomy'. *Proceedings of the Royal Society of Medicine* 29 (1936): 643–653.

———. 'Endotracheal Anesthesia'. *American Journal of Surgery* 34 (1936): 450–455.

Magill, Ivan W., Robert R. Macintosh, C. Langton Hewer, M. D. Nosworthy and W. S. McConnell. 'Lest We Forget. An Historic Meeting of the Section of Anaesthetics of the Royal Society of Medicine on 6 December 1974'. *Anaesthesia* 30 (1975): 476–490.

Magnello, Eileen. *A Centenary History of the Christie Hospital, Manchester*. Manchester: Christie Hospital NHS Trust in association with the Wellcome Unit for the History of Medicine, University of Manchester, 2001.

Marie Curie Memorial Foundation. *The Marie Curie Memorial Foundation: A Brief History 1948–1984*. London: Marie Curie Memorial Foundation, 1985.

Markle, Gerald E. and Ronald J. Troyer. 'Smoke Gets in Your Eyes: Cigarette Smoking as Deviant Behavior'. *Social Problems* 26 (1979): 611–625.

Mason, George A. 'Extirpation of the Lung'. *The Lancet* 227 (1936): 1047–1054.

Mason, Ralph C. 'The Surgical Treatment of Pulmonary Tuberculosis'. *Chest* 1 (1935): 16–17.

'Mass Miniature Radiography Service, House of Commons Debate, 2 March 1971'. *Hansard* 812 (1971): cc1666–1676.

'Mass X-ray Service, House of Commons Debate, 11 May 1970'. *Hansard* 801 (1970): c212W.

Maulitz, Russell C. *Morbid Appearances: The Anatomy of Pathology in the Early Nineteenth Century*. Cambridge: Cambridge University Press, 1987.

McKeown, Thomas. *The Role of Medicine: Dream, Mirage, or Nemesis?* London: Nuffield Trust, 1976.

Meade, Richard H. *A History of Thoracic Surgery*. Springfield, Ill: Charles C. Thomas, 1961.

Medical Research Council. 'Medical Research Council's Statement on Tobacco Smoking and Cancer of the Lung'. *Lancet* 272 (1957): 1345–1347.

———. 'Tobacco Smoking and Cancer of the Lung: Statement by the Medical Research Council'. *British Medical Journal* i (1957): 1523–1524.

Medical Research Council Working Party. 'Study of Cytotoxic Chemotherapy as an Adjuvant to Surgery in Carcinoma of the Bronchus'. *British Medical Journal* ii (1971): 421–428.

Menoret, Marie. 'The Genesis of the Notion of Stages in Oncology: The French Permanent Cancer Survey (1943–1952)'. *Social History of Medicine* 15 (2002): 291–302.

Meyer, J. A. 'Growth Rate Versus Prognosis in Resected Primary Bronchogenic Carcinomas'. *Cancer* 31 (1973): 1468–1472.

Meyer, John A. 'Surgical Resection as an Adjunct to Chemotherapy for Small Cell Carcinoma of the Lung'. In *Bronchial Carcinoma: An Integrated Approach to Diagnosis and Management*, edited by Michael Bates, 177–195. Berlin: Springer, 1984.

'Middlesex Hospital: Case of a Wound Penetrating the Cavity of the Thorax'. *The Lancet* 4 (1825): 94–95.

Miller, A. B., C. J. Baines, T. To and C. Wall. 'Canadian National Breast Screening Study: 1. Breast Cancer Detection and Death Rates Among Women Aged 40 to 49 Years'. *Canadian Medical Association Journal* 147 (1992): 1459–1476.

Miller, A. B., Wallace Fox and Ruth Tall. 'Five-year Follow-up of the Medical Research Council Comparative Trial of Surgery and Radiotherapy for the Primary Treatment of Small-Celled or Oat-Celled Carcinoma of the Bronchus. A Report to the Medical Research Council Working Party on the Evaluation of Different Methods of Therapy in Carcinoma of the Bronchus'. *Lancet* 294 (1969): 501–505.

Milroy, Robert. 'Staging of Lung Cancer'. *Chest* 133 (2008): 593–595.

Milstein, B. B. 'The Cardiothoracic Society (Pete's Club) 1959 to 1989'. *European Journal of Cardio-thoracic Surgery* 5 (1991): 339–345.

Ministry of Health – Central Health Services Council – Standing Tuberculosis Advisory Committee. *The Future of the Chest Services*. London: HMSO, 1968.

Mitchell, D. M., C. J. Emerson, J. Collyer and J. V. Collins. 'Fibreoptic Bronchoscopy: Ten Years on (Occasional Review)'. *British Medical Journal* (1980): 360–363.

Morabia, A. and F. F. Zhang. 'History of Medical Screening: From Concepts to Action'. *Postgraduate Medical Journal* 80 (2004): 463–469.

Morrison, R., T. J. Deeley and W. P. Cleland. 'The Treatment of Carcinoma of the Bronchus: A Clinical Trial to Compare Surgery and Supervoltage Radiotherapy'. *Lancet* 281 (1963): 683–684.

Mountain, Clifton F. 'A New International Staging System for Lung Cancer'. *Chest* 89 Suppl (1986): 225S–233S.

———. 'Staging of Lung Cancer: The New International System'. *Lung Cancer* 3 (1987): 4–11.

———. 'The Evolution of the Surgical Treatment of Lung Cancer'. *Chest Surgery Clinics of North America* 10 (2000): 83–104.

Mountain, Clifton F., David T. Carr and W. A. D. Anderson. 'A System for the Clinical Staging of Lung Cancer'. *American Journal of Roentgenology* 120 (1974): 130–138.

'MRC Clinical Trials Unit: Unit Profile from the MRC Network Publication Issued Spring 2005'. *MRC: Medical Research Council*, n.d. http://www.mrc.ac.uk/Ourresearch/Unitscentresinstitutes/Profiles/CTU/.

Mueller, C. Barber. *Evarts A. Graham: The Life, Lives, and Times of the Surgical Spirit of St. Louis*. Hamilton, Ontario: BC Decker, 2002.

Muggia, Franco M., Heine H. Hansen and Per Dombernowsky. 'Treatment of Small-Cell Carcinoma of Bronchus (letter)'. *Lancet* 305 (1975): 692.

Mukherjee, Siddhartha. *The Emperor of All Maladies: A Biography of Cancer*. New York: Scribner, 2010.

Murden, Jon. '"City of Change and Challenge": Liverpool Since 1945'. In *Liverpool 800: Culture, Character & History*, edited by John Belchem, 393–485. Liverpool: Liverpool University Press, 2006.

Murphy, Caroline C. S. 'A History of Radiotherapy to 1950: Cancer and Radiotherapy in Britain 1850–1950'. PhD dissertation, University of Manchester, 1986.

Murphy, Shaun. 'The Early Days of the MRC Social Medicine Research Unit'. *Social History of Medicine* 12 (1999): 389–406.

Mushin, William W. and Leslie Rendell-Baker. *The Principles of Thoracic Anaesthesia: Past and Present*. Oxford: Blackwell, 1953.

Naef, André P. 'The Mid-century Revolution in Thoracic and Cardiovascular Surgery: Parts 1–6'. *Interactive Cardiovascular and Thoracic Surgery* 2 (2003): 219–226, 431–449, and 3 (2004): 3–10, 213–222, 415–422, 535–541.

———. *The Story of Thoracic Surgery: Milestones and Pioneers*. Toronto etc: Hogrefe & Huber, 1990.

Nash, F. A., J. M. Morgan and J. G. Tomkis. 'South London Lung Cancer Study'. *British Medical Journal* ii (1968): 715–721.

Nathoo, Ayesha. *Hearts Exposed: Transplants and the Media in 1960s Britain*. Basingstoke: Palgrave Macmillan, 2009.

Newsholme, Arthur. 'The Statistics of Cancer'. *The Practitioner* (April 1899): 371–384.

Nicholson, W. Frank, Miles Fox and A. Graham Bryce. 'Review of 910 Cases of Bronchial Carcinoma'. *Lancet* 269 (1957): 296–298.

Nicolson, Malcolm and George W. Lowis. 'The Early History of the Multiple Sclerosis Society of Great Britain and Northern Ireland: A Socio-historical Study of Lay/practitioner Interaction in the Context of a Medical Charity'. *Medical History* 46 (2002): 141–174.

'No Smoke Without Harm'. *Economist* (18 January 1964): 205.

'No Smoking?' *Economist* (10 March 1962): 878–879.

Nosworthy, M. D. 'Anaesthesia in Chest Surgery, with Special Reference to Controlled Respiration and Cyclopropane'. *Proceedings of the Royal Society of Medicine* 34 (1941): 479–506.

NY-ELCAP investigators. 'Tobacco Money: Up in Smoke? [letter]'. *Lancet* 360 (2002): 1980–1981.

'Obituary Notices: D. M. Kissen'. *British Medical Journal* i (1968): 773.

'Obituary Notices: Gwen Hilton'. *British Medical Journal* iii (1971): 253.

'Obituary: Arthur Tudor Edwards'. *Lancet* 248 (1946): 365–366.

'Obituary: Hugh Morriston Davies'. *Lancet* 285 (1965): 387–389.

'Obituary: J. E. H. Roberts'. *Thorax* 3 (1948): 185–187.

'Obituary: James Ernest Helme Roberts'. *The Lancet* 252 (1948): 398–399.

'Obituary: Mr Richard Dimbleby: Professionalism in Broadcasting'. *The Times*, 23 December 1965.

'Obituary: Professor Norman Bleehen 1930–2008'. *IASLC Newsletter* (December 2008).

'Obituary: Roy Castle'. *The Times*, 3 September 1994.

Ochsner, Alton. *Smoking and Cancer: A Doctor's Report*. New York: Julian Messner, 1954.

Olch, Peter D. 'Evarts A. Graham in World War I: The Empyema Commission and Service in the American Expeditionary Forces'. *Journal of the History of Medicine and Allied Sciences* 44 (1989): 430–446.

Olsen, Ole and Peter C. Gøtzsche. 'Cochrane Review on Screening for Breast Cancer with Mammography'. *The Lancet* 358 (2001): 1340–1342.

'One Man's Smoke'. *Economist* (19 July 1969): 56–57.

Onuigbo, W. I. B. 'Lung Cancer in the Nineteenth Century'. *Medical History* 3 (1959): 69–77.

Otis, Laura. *Müller's Lab*. Oxford & New York: Oxford University Press, 2007.

Pack, George T. and Irving M. Ariel. 'A Half Century of Effort to Control Cancer; an Appraisal of the Problem and an Estimation of Accomplishments'. In *Fifty*

Years of Surgical Progress 1905–1955, edited by Loyal Davis, 59–161. Chicago: Franklin H. Martin Memorial Foundation, 1955.

Palladino, Paolo. 'Discourses of Smoking, Health, and the Just Society: Yesterday, Today, and the Return of the Same?' *Social History of Medicine* 14 (2001): 313–335.

Palmer, J. W. 'Smoking, Caning, and Delinquency in a Secondary Modern School'. *British Journal of Preventive and Social Medicine* 19 (1965): 18–23.

Paneth, Matthias. 'The Brompton Hospital and Cardiothoracic Surgery'. *European Journal of Cardio-thoracic Surgery* 5 (1991): 6–12.

Parascandola, Mark. 'Cigarettes and the US Public Health Service in the 1950s'. *American Journal of Public Health* 91 (2001): 196–205.

Parascandola, Mark, Douglas L. Weed and Abhijit Dasgupta. 'Two Surgeon General's Reports on Smoking and Cancer: A Historical Investigation of the Practice of Causal Inference'. *Emerging Themes in Epidemiology* 3 (2006): 1.

Parker-Pope, Tara. 'Cancer Funding: Does It Add Up?' *Well: Tara Parker-Pope on Health*, 6 March 2008. http://well.blogs.nytimes.com/2008/03/06/cancer-funding-does-it-add-up/.

Pässler, Hans. 'Ueber das primäre Carcinom der Lunge'. *Virchows Archiv* 145 (1896): 191–278.

Pausch, Randy. *The Last Lecture*. London: Hodder & Stoughton, 2008.

Payne, Sarah. '"Smoke Like Man, Die Like a Man"?: A Review of the Relationship Between Gender, Sex and Lung Cancer'. *Social Science and Medicine* 53 (2001): 1067–1080.

Peake, Mike, Paul Beckett, Ian Woolhouse and Roz Stanley. *National Lung Cancer Audit Report 2011*. http://www.ic.nhs.uk/canceraudits/lung.

Peitzman, Steven J. *Dropsy, Dialysis, Transplant: A Short History of Failing Kidneys*. Baltimore: Johns Hopkins University Press, 2007.

Peters, Matthew J. 'Should Smokers Be Refused Surgery? Yes'. *British Medical Journal* 334 (2007): 20.

Pett, S. B., Jr, J. A. Wernly and B. F. Akl. 'Lung Cancer – Current Concepts and Controversies'. *The Western Journal of Medicine* 145 (1986): 52–64.

Pickstone, John V. 'Contested Cumulations: Configurations of Cancer Treatments through the Twentieth Century'. *Bulletin of the History of Medicine* 81 (2007): 164–196.

———. *Medicine and Industrial Society: A History of Hospital Development in Manchester and Its Region 1725–1946*. Manchester: Manchester University Press, 1985.

Pinell, Patrice. 'Cancer'. In *Medicine in the Twentieth Century*, edited by Roger Cooter and J. V. Pickstone, 671–686. Amsterdam: Harwood Academic, 2000.

Pitt, G. N. 'Malignant Disease of Bronchial Glands'. *Transactions of the Pathological Society London* 39 (1888): 54–56.

'Plain or Filter?' *Economist* (16 April 1960): 265–266.

Plaut, Alfred. 'Rudolf Virchow and Today's Physicians and Scientists'. *Bulletin of the History of Medicine* 27 (1953): 236–251.

Posner, E., L. A. McDowell and K. W. Cross. 'Mass Radiography and Cancer of the Lung'. *British Medical Journal* i (1959): 1213–1218.

———. 'Place of Mass Radiography in Relation to Lung Cancer in Men'. *British Medical Journal* ii (1963): 1156–1160.

Powell, Kenneth. 'Obituary: Maggie Jencks'. *The Independent*, 14 July 1995.

Prochaska, Frank K. *Philanthropy and the Hospitals of London: The King's Fund 1897–1990*. Oxford: Clarendon Press, 1992.

Proctor, Robert N. 'Commentary: Schairer and Schöniger's Forgotten Tobacco Epidemiology and the Nazi Quest for Racial Purity'. *International Journal of Epidemiology* 30 (2001): 31–34.

———. *The Nazi War on Cancer*. Princeton: Princeton University Press, 1999.

Pyke, D. A. 'Cigarette Smoking and Bronchial Carcinoma: Effect of the Association Upon Smoking Habits of a Group of Doctors'. *British Medical Journal* i (1955): 1115–1116.

'Radiotherapy and Bronchial Carcinoma (editorial)'. *Lancet* 262 (1953): 1298–1299.

'Radiotherapy for Lung Cancer (editorial)'. *Lancet* 266 (1955): 963.

Rather, L. J. *The Genesis of Cancer: A Study in the History of Ideas*. Baltimore & London: Johns Hopkins University Press, 1978.

Reid, D. D., A. M. Adelstein and W. P. D. Logan. 'Percy Stocks: An Appreciation'. *British Journal of Preventive and Social Medicine* 29 (1975): 65–72.

Robinson, Samuel. 'The Present and Future in Thoracic Surgery'. *Archives of Surgery* 6 (1923): 247–255.

Rosenberg, Charles E. and Janet Golden, eds. *Framing Disease: Studies in Cultural History*. New Brunswick, N.J: Rutgers University Press, 1992.

Rosenblatt, Milton B. 'Lung Cancer in the 19th Century'. *Bulletin of the History of Medicine* 38 (1964): 395–425.

Ross, Walter. *Crusade: The Official History of the American Cancer Society*. New York: Arbor House, 1987.

Rossi, Paul N. *Fighting Cancer with More Than Medicine: A History of Macmillan Cancer Support*. Stroud, Gloucestershire: History Press, 2009.

Rothstein, William G. *Public Health and the Risk Factor: A History of an Uneven Medical Revolution*. Rochester, NY: University of Rochester Press, 2003.

Royal College of Physicians. *Smoking and Health Now: A New Report and Summary on Smoking and Its Effects on Health from The Royal College of Physicians of London*. London: Pitman Medical Publishing, 1971.

———. *Smoking and Health: A Report of The Royal College of Physicians on Smoking in Relation to Cancer of the Lung and Other Diseases*. London: Pitman Medical Publishing, 1962.

Rubin, Sanford A. 'Lung Cancer: Past, Present, and Future'. *Journal of Thoracic Imaging* 7 (1991): 1–8.

Samet, Jonathan M., Erika Avila-Tang, Paolo Boffetta, Lindsay M. Hannan, Susan Olivo-Marston, Michael J. Thun and Charles M. Rudin. 'Lung Cancer in Never Smokers: Clinical Epidemiology and Environmental Risk Factors'. *Clinical Cancer Research* 15 (2009): 5626–5645.

Sauerbruch, Ferdinand. 'Zur Pathologie des offenen Pneumothorax und die Grundlagen meines Verfahrens zu seiner Ausschaltung'. *Mitteilungen aus den Grenzgebieten der Medizin und Chirurgie* 13 (1904): 399–482.

Saxon, Wolfgang. 'Ernst Wynder, 77, a Cancer Researcher, Dies'. *New York Times*, 16 July 1999.

Scadding, J. G. 'Treatment of Bronchial Carcinoma (letter)'. *Lancet* 289 (1967): 157.

Scadding, J. G., John R. Bignall, L. G. Blair, W. P. Cleland, A. L. D'Abreu, Wallace Fox, D. A. G. Galton, J. Gough, Alexander Haddow, J. F. D'Arcy Hart, J. F. Heffernan, Austin Bradford Hill, A. M. Jelliffe, A. B. Miller, I. Sutherland and Brian Windeyer. 'Comparative Trial of Surgery and Radiotherapy for the

Primary Treatment of Small-Celled or Oat-Celled Carcinoma of the Bronchus: First Report to the Medical Research Council by the Working-Party on the Evaluation of Different Methods of Therapy in Carcinoma of the Bronchus'. *Lancet* 288 (1966): 979–986.

Scannell, J. Gordon. 'Historical Perspectives of the American Association for Thoracic Surgery: Samuel J. Meltzer (1851–1920)'. *Journal of Thoracic and Cardiovascular Surgery* 111 (1996): 905–906.

———. 'Historical Perspectives of the American Association for Thoracic Surgery: Samuel Robinson (1875–1947)'. *Journal of Thoracic and Cardiovascular Surgery* 112 (1996): 562.

Schairer, E. and E. Schöniger. 'Lung Cancer and Tobacco Consumption'. *International Journal of Epidemiology* 30 (2001): 24–27.

'School 1963 – One in Five Are Smokers'. *Daily Mirror*, 16 February 1963.

Schüttmann, Werner. 'Beitrag zur Geschichte der Schneeberger Lungenkrankheit, des Strahlenkrebses der Lunge durch Radon und seine Folgeprodukte'. *NTM* 25 (1988): 83–96.

———. 'Schneeberg Lung Disease and Uranium Mining in the Saxon Ore Mountains (Erzgebirge)'. *American Journal of Industrial Medicine* 23 (1993): 355–368.

'Section of Radiology: Discussion on the Place of Miniature Radiography in the Diagnosis of Diseases of the Chest'. *Proceedings of the Royal Society of Medicine* 36 (1942): 155–160.

Shapiro, Sam, W. Venet, P. Strax and L. Venet. *Periodic Screening for Breast Cancer: The Health Insurance Plan Project and Its Sequelae, 1963–1986*. Baltimore: Johns Hopkins University Press, 1988.

'Shock Plan to Ban Smoking in Public'. *Daily Mirror*, 10 December 1966.

Simon, G. 'Radiographic Aspects of Carcinoma of the Lung'. In *Some Aspects of Carcinoma of the Bronchus and Other Malignant Diseases of the Lung. A Symposium Held at King Edward VII Hospital, Midhurst, July 4th and 5th, 1966*, 87–94. Midhurst: King Edward VII Hospital, 1966.

'Sir David Smithers: Obituary'. *The Times*, 29 July 1995.

Sitsen, A. E. 'Über die Häufigkeit des Lungenkrebses'. *Zeitschrift für Krebsforschung* 36 (1932): 313–318.

———. 'Wird der Lungenkrebs häufiger? (Eine kritisch-statistische Erörterung)'. *Zeitschrift für Krebsforschung* 42 (1935): 30–45.

Slaughter, Frank G. *The New Science of Surgery*. New York: J. Messner, 1946.

Smart, Joseph and Gwen Hilton. 'Radiotherapy of Cancer of the Lung: Results in a Selected Group of Cases'. *Lancet* 267 (1956): 880–881.

Smith, George Davey, Sabine Strobele and Matthias Egger. 'Smoking and Death [letter]'. *British Medical Journal* 310 (1995): 396.

Smithers, David W. 'Clinical Cancer Research'. *Lancet* 267 (1956): 253–257.

———. 'Facts and Fancies About Cancer of the Lung'. *British Medical Journal* i (1953): 1235–1239.

———. *The X-ray Treatment of Accessible Cancer*. London: Arnold, 1946.

Smithers, David W., Katherine M. H. Branson and Herman O. Hartley. *The Royal Cancer Hospital Mechanically Sorted Punched Card Index System*. London: Royal Marsden Hospital, 1946.

Smithers, David Waldron. *Not a Moment to Lose: Some Reminiscences*. Cambridge: The Memoir Club; BMJ, 1989.

'Smoke Abatement'. *Economist* (6 July 1957): 15–16.

'Smokers Carry on as Usual'. *Daily Mirror*, 14 January 1964.

'Smokers' Kids "Are Yobs"'. *The Sun*, 31 July 2007.

'Smoking and Cancer'. *Economist* (20 February 1954): 523–524.

'Smoking and Cancer'. *The Times*, 13 February 1954.

'Smoking and Health: 4 New Moves Are Forecast. All Out Research'. *Daily Mirror*, 13 February 1954.

Smyth, John, ed. *The Management of Lung Cancer*. London: Edward Arnold, 1984.

Snowdon, Christopher. *Velvet Glove, Iron Fist: A History of Anti-Smoking*. London: Little Dice, 2009.

Sobin, Leslie H. 'International Histological Classification of Tumors, Second Edition'. *Cancer* 63 (1989): 907.

Sobin, Leslie H. and Harvey W. Baker. 'Request for Suggestions on a Unified System for Classifying and Staging Cancer'. *CA: A Cancer Journal for Clinicians* 34 (1984): 304.

Spiro, S. G. 'Diagnosis and Staging'. In *Lung Cancer*, edited by William Duncan, 16–29. Berlin: Springer, 1984.

Spiro, Stephen G., Nichole T. Tanner, Gerard A. Silvestri, Sam M. Janes, Eric Lim, Johan F. Vansteenkiste and Robert Pirker. 'Lung Cancer: Progress in Diagnosis, Staging and Therapy'. *Respirology* 15 (2010): 44–50.

Sporn, M. B. 'The War on Cancer'. *The Lancet* 347 (1996): 1377–1381.

Stapleton, John. 'Cigarette Smoking Prevalence, Cessation and Relapse'. *Statistical Methods in Medical Research* 7 (1998): 187–203.

Stokes, William. *A Treatise on the Diagnosis and Treatment of Diseases of the Chest. Part 1. Diseases of the Lung and Windpipe*. Dublin: Hodges and Smith, 1837.

Stolley, Paul D. and Tamar Lasky. *Investigating Disease Patterns: The Science of Epidemiology*. New York: Scientific American Library, 1995.

Stott, H., R. J. Stephens, W. Fox and D. C. Roy. '5-year Follow-up of Cytotoxic Chemotherapy as an Adjuvant to Surgery in Carcinoma of the Bronchus'. *British Journal of Cancer* 34 (1976): 167–173.

Sullum, Jacob. *For Your Own Good: The Anti-smoking Crusade and the Tyranny of Public Health*. New York: The Free Press, 1998.

Szabo, Jason. *Incurable and Intolerable: Chronic Disease and Slow Death in Nineteenth-century France*. New Brunswick, N.J: Rutgers University Press, 2009.

Talley, Colin, Howard I. Kushner and Claire Sterk. 'Lung Cancer, Chronic Disease Epidemiology and Medicine, 1948–1964'. *Journal of the History of Medicine and Allied Sciences* 59 (2004): 329–374.

Taylor, John. 'Clinical Lecture, Delivered at University College Hospital: Cancer of the Right Lung, Vertebral Column, Sternoclavicular Articulation, Stomach and Kidneys'. *Lancet* 37 (1842): 873–904.

Taylor, Peter. *Smoke Ring: The Politics of Tobacco*. London: Bodley Head, 1984.

'Teen Behaviour "Better Than 1985"'. *BBC News*, 18 May 2006. http://news.bbc.co.uk/1/hi/uk/4994538.stm.

'Terry Kavanagh – Conquering Lung Cancer – I Can Get Well'. *Cancer Active*, 14 August 2012.

Thatcher, N., M. Ranson, S. M. Lee, R. Niven and H. Anderson. 'Chemotherapy in Non-small Cell Lung Cancer'. *Annals of Oncology* 6, Suppl 1 (1995): 83–94; discussion 94–95.

Thatcher, Nick and Stephen Spiro. *New Perspectives in Lung Cancer*. London: BMJ Publishing Group, 1994.

'The Late Dr Warren'. *The Lancet* 25 (1836): 550–552.

'The Man Who Worried: Suicide after the "Smokes" Warning'. *Daily Mirror*, 13 March 1963.

Timmermann, Carsten. 'As Depressing as It Was Predictable? Lung Cancer, Clinical Trials, and the Medical Research Council in Postwar Britain'. *Bulletin of the History of Medicine* 81 (2007): 312–334.

———. '"Just Give Me the Best Quality of Life Questionnaire"': The Karnofsky Scale and the History of Quality of Life Measurements in Cancer Trials'. *Chronic Illness* 9 (2013): 179–190.

———. 'Running Out of Options: Surgery, Hope and Progress in the Management of Lung Cancer, 1950s to 1990s'. In *Cancer Patients, Cancer Pathways: Historical and Sociological Perspectives*, edited by Carsten Timmermann and Elizabeth Toon, 36–56. Basingstoke: Palgrave Macmillan, 2012.

Tod, Margaret. *An Inquiry into the Extent to Which Cancer Patients in Great Britain Receive Radiotherapy*. Altrincham: John Sherratt, 1949.

Todd, G. F. *Social Class Variations in Cigarette Smoking and Mortality from Associated Diseases*. London: Tobacco Research Council, 1976.

'Tuberculosis and the National Health Service: Report of a B.M.A. Group Committee'. *British Medical Journal* ii (1950): 1382–1385.

'Tudor Edwards Memorial'. *Lancet* 249 (1947): 809–810.

'Tudor Edwards Memorial'. *Lancet* 252 (1948): 788.

Tweedale, Geoffrey. 'Asbestos and Its Lethal Legacy'. *Nature Reviews. Cancer* 2 (2002): 311–315.

———. *Magic Mineral to Killer Dust: Turner & Newall and the Asbestos Hazard*. Oxford: Oxford University Press, 2000.

Twycross, Robert G. 'The Terminal Care of Patients with Lung Cancer'. *Postgraduate Medical Journal* 49 (1973): 732–737.

Ueyama, Takahiro and Christophe Lecuyer. 'Building Science-based Medicine at Stanford: Henry Kaplan and the Medical Linear Accelerator, 1948–1975'. In *Devices and Designs: Medical Technologies in Historical Perspective*, edited by Carsten Timmermann and Julie Anderson, 137–155. Basingstoke: Palgrave Macmillan, 2006.

Uhlig, Margarete. 'Über den Schneeberger Lungenkrebs'. *Virchows Archiv* 230 (1921): 76–98.

Valier, Helen and Carsten Timmermann. 'Clinical Trials and the Reorganization of Medical Research in Post-Second World War Britain'. *Medical History* 52 (2008): 493–510.

Vayena, Eftychia. 'Cancer Detectors: An International History of the Pap Test and Cervical Cancer Screening, 1928–1970'. PhD dissertation, University of Minnesota, 1999.

Virchow, Rudolf. *Die Cellularpathologie in ihrer Begründung auf physiologische und pathologische Gewebelehre*. Berlin: A. Hirschwald, 1858.

———. *Die krankhaften Geschwülste: Dreissig Vorlesungen, gehalten während des Wintersemesters 1862–1863 an der Universität zu Berlin*. Berlin: A. Hirschwald, 1863.

Viscusi, W. Kip. *Smoking: Making the Risky Decision*. New York: Oxford University Press, 1992.

Walloch, Jami. 'Pulmonary Cytopathology in Historical Perspective'. In *Lung Cancer: The Evolution of Concepts. Vol. 1*, edited by John G. Gruhn and Steven T. Rosen, 169–195. New York: Field & Wood Medical Publishers, 1989.

Walshe, Walter Hayle. *The Nature and Treatment of Cancer*. London: Taylor & Walton, 1846.

Ward. 'Seamen's Hospital, "Dreadnought": Medullary Cancer of Mediastinal Glands and Left Lung'. *Lancet* 89 (1867): 238–239.

Watson, W. L. and J. W. Berg. 'Oat Cell Lung Cancer'. *Cancer* 15 (1962): 759–768.

Watson, W. L. 'Thoracic Surgical Service 1940–1941'. In *Annual Report 1940–1942*. New York: Memorial Cancer Hospital, 1943.

Webb, Kathleen. *Hugh Morriston Davies: Pioneer Thoracic Surgeon, 1879–1965*. Ruthin: Coelion Trust, 1998.

Webster, Charles. 'Tobacco Smoking Addiction: A Challenge to the National Health Service'. *British Journal of Addiction* 79 (1984): 7–16.

Welch, H. Gilbert. *Should I Be Tested for Cancer? Maybe Not and Here's Why*. Berkeley: University of California Press, 2004.

Welch, H. Gilbert, Steven Woloshin and Lisa M. Schwartz. 'How Two Studies on Cancer Screening Led to Two Results'. *New York Times*, 13 March 2007.

Wheeler-Bennett, John W. *King George VI: His Life and Reign*. London: Macmillan, 1958.

Wilkes, Eric. 'Terminal Cancer at Home'. *Lancet* 285 (1965): 799–801.

Williams, W. Roger. *The Natural History of Cancer: With Special Reference to Its Causation and Prevention*. London: William Heinemann, 1908.

Wilson, J. Walter. 'Virchow's Contribution to the Cell Theory'. *Journal of the History of Medicine and Allied Sciences* 2 (1947): 163–178.

Wiltshaw, Eve. *A History of the Royal Marsden Hospital*. Edgeware: Altman, 1998.

Winter, B. 'Early Symptomatology of Carcinoma of the Lung'. *Canadian Medical Association Journal* 63 (1950): 134–138.

Wishart, Adam. *One in Three: A Son's Journey into the History and Science of Cancer*. London: Profile, 2006.

Wolff, Jacob. *Die Lehre von der Krebskrankheit: Von den ältesten Zeiten bis zur Gegenwart*. Jena: Gustav Fischer, 1907.

Woloshin, Steven, Lisa M. Schwartz and H. Gilbert Welch. 'Tobacco Money: Up in Smoke?' *Lancet* 359 (2002): 2108–2111.

World Health Organization. 'Fact Sheet No 297: Cancer'. *WHO Media Centre*. http://www.who.int/mediacentre/factsheets/fs297/en/.

———. 'General Preface to the Series'. In *Histological Typing of Lung Tumours*, 9–10. 2nd ed. International Histological Classification of Tumours 1. Geneva: World Health Organization, 1981.

———. *Histological Typing of Lung Tumours*. 2nd ed. International Histological Classification of Tumours 1. Geneva: World Health Organization, 1981.

———. *Manual of the International Statistical Classification of Diseases, Injuries and Causes of Death*. 1975 Revision. Geneva: World Health Organization, 1977.

Wynder, E L. 'Tobacco as a Cause of Lung Cancer: Some Reflections'. *American Journal of Epidemiology* 146 (1997): 687–694.

Wynder, Ernst L. and Evarts A. Graham. 'Tobacco Smoking as Possible Etiologic Factor in Bronchiogenic Carcinoma'. *Journal of the American Medical Association* 143 (1950): 329–336.

'X-ray Units, House of Commons Debate, 30 Jan 1978'. *Hansard* 943 (1978): c51W.

Yoshioka, A. 'Use of Randomisation in the Medical Research Council's Clinical Trial of Streptomycin in Pulmonary Tuberculosis in the 1940s'. *British Medical Journal* 317 (1998): 1220–1223.

Yoshioka, Alan. 'Streptomycin in Postwar Britain: A Cultural History of a Miracle Drug'. *Clio Medica* 66 (2002): 203–227.

Zimmermann, Susanne, Matthias Egger and Uwe Hossfeld. 'Commentary: Pioneering Research into Smoking and Health in Nazi Germany – The "Wissenschaftliches Institut zur Erforschung der Tabakgefahren" in Jena'. *International Journal of Epidemiology* 30 (2001): 35–37.

Index

Abbey Smith, Roger, 52, 110
accidental productions (according to
    Laennec), 15–16
  analogous and non-analogous
    productions, 16
acetate of lead (traditional remedy), 1
Ackerknecht, Erwin, 23
Action on Smoking and Health (ASH),
    118, 123–6, 136, 155
Addenbrooke's Hospital, Cambridge,
    135, 150
addiction, 152
Adler, Isaac, 11, 24, 25, 31–3, 34
Alder Hey Children's Hospital,
    Liverpool, 158
Alliance for Lung Cancer Advocacy,
    Support and Education, 149
Allison, P. R., 114
American Association for Cancer
    Research, 166
Scientist/Survivor Programme, 166
American Association for Thoracic
    Surgery, 42, 53
American Cancer Society, 76, 77, 141,
    157, 168
American College of Radiology, 141
American College of Surgeons, 141
American Joint Committee on Cancer
    (AJC), 141
American Joint Committee on Cancer
    Staging and End Results
    Reporting, 141
Task Force on Lung Cancer, 141
anaesthesia, 34, 37, 40–1
  importance for chest surgery, 49
  one-lung, for lobectomies and
    pneumonectomies, 41
Andral, Gabriel, 20, 26
antibiotics, 34, 47, 52
antisepsis, 36
anti-smoking activism, 155–6
Armitage, P., 106
Aronowitz, Robert, 174

artificial pneumothorax, 35
asbestos, 91
asepsis, 36
Ash, D., 145
ASH see Action on Smoking and
    Health
Astel, Karl, 76–7
Atkinson Morley Hospital, London,
    60
Atomic Energy Commission, 74
auscultation (diagnostic procedure),
    13, 14–16, 18, 19, 21
  see also percussion; physical
    examination

Baguley Hospital, Manchester, 57–8
Ball, Keith, 124–5, 136
Barnes, Emm, 156
Barnes Hospital, St Louis, 48–9, 75–6
Barett, Norman R., 113–14
barium-swallow, 142
Bayer, Ronald, 156
Bayle, Gaspard-Laurent, 15, 20
belladonna plasters, 21
Belcher, John Rashley (Jack), 110,
    114, 116–17
Benbow, Mary, 1–4, 20
Berridge, Virginia, 123, 125
Bichat, M. F. Xavier, 14
Bignall, John Reginald, 100–1, 104,
    115, 119, 129–30, 173
  Carcinoma of the Lung (1958), 100
biopsy, 3, 48
Birmingham Regional Hospital Board,
    112
Births and Deaths Registration Act
    (1874), 30
Bleehen, Norman M., 132–6, 143, 144
bone sarcoma, 102–3
Branson, Catherine, 60
breast cancer, 26, 33, 34, 173–4
  culture of survivorship, 156, 173–4
  genes, 166

as a model cancer, 34
as a target for surgery, 26, 33, 34
Brett, G. Z., 129
Bright, Richard, 1–2
British Council, 61
British and Foreign Medical Review: A Quarterly Journal of Practical Medicine, 17
British Empire Cancer Campaign (BECC), 104, 133
    Clinical Research Committee, 71, 74
    survey of cancer in London, 96–7, 117
British Lung Foundation (BLF), 158–9
Broadgreen Hospital, Liverpool, 158
Brompton Cocktail, 121
Brompton Hospital (for Consumptives and Diseases of the Chest), 13, 17, 22, 36, 40, 49–53, 59, 60, 61, 94, 96–101, 115, 116, 134, 141–3, 165
bronchiectasis, 34, 35, 39, 44, 47
    surgical treatment of, 42–4
bronchitis, 16, 45, 69–70
bronchoscopy, 3, 44, 48, 51, 94, 97
    introduction of flexible, fibreoptic bronchoscope, 115
Bonser, Georgiana, 65–6
Burchenal, Joseph, 62
Burnet, Neil, 150–1
Bryce, Alexander Graham, 53, 58

Cade, Stanford, 60
Cambridge University, Department of Clinical Oncology, 135
Cambrosio, Alberto, 119, 132
Campbell, D. F., 102
cancer
    and basic research, 157
    of the bladder, 102–3, 105
    classification *see* lung cancer, tumours
    clinical research, 132
    as a constitutional disease, 29
        see also dyscrasias
    deaths, 157
    diagnosis, 60
    as a disease of cells, 2, 13–14, 22–6

as a disease of tissues, 2, 13–14
    and epidemiology, 29–31
    hospitals, 157
    and humoral pathology, 23–4
    increase in incidence, 30, 31
    local or systemic, 16, 27–8, 29
    mortality figures, 6–7, 30
    of the oesophagus, 102–3
    panic (cancerphobia), 88–9
    reduction of mortality, 4
    reorganisation of services, 58–61
    research, 175
    research funding, 6, 149–51, 165
    and specialization, 135
    staging, 139–43
        see also lung cancer
    stigma, 148
    treatment, 58–61
        see also chemotherapy; radiotherapy; surgery
    treatment outcomes, 4, 5
        see also carcinoma; sarcoma; tumours
Cancer Research Campaign (CRC), 135, 137, 143, 144–5, 157
Cancer Research UK, 157
carcinoma, 25, 26
    distinction from sarcoma, 25
Carcinoma of the Lung (John R. Bignall, 1958), 100
Carmichael, Jim, 138
Carr, David, 141–3
Carswell, Robert, 20
Carter, R. L., 145
Cartwright, Anne, 87–90
Castle, Fiona, 164
Castle, Roy, 159, 163–4
    Roy Castle Cause for Hope Appeal, 164
    Roy Castle Foundation, 147, 157–70
    Roy Castle Tour of Hope, 164
causes of death, 6
    certification, 30
cell theory, 23
cellular pathology, 13–14, 22–6, 28, 91
Cellularpathologie, Die (Rudolf Virchow 1858), 23

census data, 30
cerebriform tumours, 15–16
Chapple, Alison, 149
charities, 147, 156–8, 165–6
Charlie's Club, 116
Chelsea Hospital for Women, 60
chemoprevention, 167
chemotherapy, 3, 5, 7, 62, 103
chest centres *see* chest units
chest surgeons
    attitudes towards progress, 35, 36,
        52, 53–4, 114
    experiments on animals, 35, 48, 52
    numbers of consultants and senior
        registrars, 35, 56, 58
    training, 55
    see also American Association for
        Thoracic Surgery; Society of
        Thoracic Surgeons
chest surgery, 7–8, 34–63
    in Britain prior to the National
        Health Service (NHS), 49–58
    descriptions of operations, 42–3,
        44, 45, 47–9
    documentary film, 'Surgery in
        Chest Disease' (1943), 61–2
    exploratory, 46
    importance of speed, 39, 43
    increase in number of operations,
        51–2
    mortality, 44–5, 47, 49, 120
    and the National Health Service
        (NHS), 35, 56
    and pulmonary physiology, 34,
        53–4
    risks to patients, 42–5, 49, 51
    in sanatoria, 40
    specialization, 50–6
    see also anaesthesia; artificial
        pneumothorax; collapse
        therapy; *Journal of Thoracic
        Surgery*, lobectomy;
        pneumonectomy;
        thoracoplasty; thoracotomy
chest units, 3, 34, 35, 54–8
    in provincial hospitals, 56–8
Chester Beattie Institute, 59, 74, 134
childhood cancers, 4, 156–7, 174–5
cholera, 29

Christian, Sheila, 158–9
Christie Hospital, Manchester, 58, 59,
    60, 101, 134, 137–8, 149
Churchill Hospital, Headington, 134
civic registration, 30
Clarke, Kenneth, 128
classification, 14–16, 22–5, 29,
    139–43, 153
    see also lung cancer
Cleland, William Paton, 100
clinical observation, 14
    see also diagnosis; lung cancer,
        diagnosis; physical
        examination
clinical record keeping, 59–60, 99
clinical research, 99, 103, 132, 157
    Clinical Cancer Research (article by
        David Smithers, 1956), 59
    doubts about its value, 136
clinical science tradition (associated
    with Thomas Lewis), 59
clinical trials, 132
    ethics, 104–5, 136
    protocols, 137, 139
    see also randomized controlled trials
collapse therapy, 35
College of American Pathologists, 141
compound microscope, 23
compresses, hot, 21
computed tomography (CT), 131,
    141–2, 167–8
Connaught Hospital, Walthamstow,
    116
Connors, T. A., 134
consumerism, 125
consumption (illness), 2, 6, 14, 15, 18
    transformation of understanding,
        14
Coordinating Committee on Cancer
    Research (CCCR), 144
Cooter, Roger, 52
coughing up blood, 1, 3, 19
Cowdry, Edward, 76
Crafoord, Clarence *On the Technique
    of Pneumonectomy in Man* (1938),
    49
Craig, Frank, 3–4
Craig, Mary, 3–4
Cross, K. W., 112

Crossman, Richard (Dick), 85
Crowther, Derek, 137–8
Cyclopaedia of Practical Medicine, 17

D'Abreu, Alphonsus, 102, 104, 106
D'Arcy Hart, Philip, 107
Daley, Allen, 69
Dark, John, 58
Daube, Mike, 123, 125
Davidson, Maurice, 51, 99
Davies, Anne, 5, 20–2
Davies, Hugh Morriston, 36–7, 38–40,
    53, 56, 158
  accident, 39
  lobectomy technique, 39, 44, 45
  operating lung cancer patient, 45–6
  Positive Pressure Anaesthetic
    Machine, 38–9 *Figure 3.2*
  priority claims, 50
death certificates, 6, 71
DeBakey, Michael, 75
Deeley, Thomas J., 134
Denoix, Pierre, 140–1
diagnosis, differential, 18–22
  see also lung cancer
Dictionaire des Sciences Médicales
    (1815), 15
digitalis, 1
Dimbleby Cancer Care, 163, 165–6
Dimbleby, Richard, 163
disability, 148
disposition, 16
Dodd, Ken, 159
Doll, Richard, 73–5, 77–9, 80, 133,
    155–6
Donaldson, Malcolm, 71, 74
Donnelly, Raymond (Ray), 157–69,
    173
Dreadnought Seamen's Hospital, 25
Duffin, Jacalyn, 15–16
Duguid, John Bright, 65–6
dying, 100–1, 121–3, 175
dyscrasias, 23–4

early detection, 7, 46, 100, 112,
    126–32, 166–9
  futility of, 114–15
Early Lung Cancer Action Project
    (ELCAP), 167–8

East Anglian Cancer Register, 150
Edwards, Arthur Tudor, 50–3, 61
Ehrenreich, Barbara, 174
Elliott, Rosemary, 151, 155
Elton, Arthur, 61
Emergency Medical Service (EMS), 35,
    54, 56
Emperor of all Maladies, The
    (Mukherjee 2010), 8
emphysema, 16
empyema, 18, 19, 34
  following influenza, 42
Empyema Commission, 42
encephaloids (tumours), 15–16, 18,
    19, 22, 24, 25–6
Enquête Permanente Cancer
    (Permanent Cancer Survey), 140
epidemiology, 29–31, 65–78, 91, 169
  case control study by Doll and Hill,
    72–5
  cohort studies by Doll and Hill, and
    Hammond, 77–8
  comparison with laboratory
    experiments, 78
Evans, Horace, 94
Everton Football Club, 160
Eysenck, Hans J., 152–3

Fairley, Gordon Hamilton, 133, 137,
    144
Farr, William, 29–30
fever (as a diagnosis), 12
Field, John, 166–9
Fisher, R. A., 152
First World War
  chest wounds, 41–2
  effect on chest surgery, 35, 41–3
  innovation in anaesthesia, 40–1
fistulae, 48, 52
Fletcher, Charles, 82, 89, 124
flow charts representing clinical trial
    protocols, 136–7
Forbes, John, 16–18
  translation of R. T. H. Laennec's
    traité de l'auscultation into
    English, 16–17
Forlanini, Carlo, 36
Forrest, Patrick, 128
Fowler, James Kingston, 36

Fox, Maurice, 173
Fox, Wallace, 134
Frankfurt (Main), cancer incidence
    figures, 31
Fry, John (GP), 69–70

Galton Laboratory, UCL, 71
Gaudillière, Jean-Paul, 157
General Practitioners (GPs), 69–70
General Register Office (GRO), 29–31,
    71, 74
George VI, 3, 93–6
German Pathological Society, 65
Glantz, Leonard, 151
gloom, 113–17
Godber, George, 124
Goffman, Erving, 147–8
Goldstraw, Peter, 142–3, 145
Gough, Jethro, 102, 104
Graham, Evarts, 42, 46–9, 53–4, 98
    involvement with Empyema
        Commission, 42
    pioneering pulmonary resection,
        46–7, 48–9
    smoking habits, 49
    work with Ernst Wynder, 75–6,
        77–8
Green, Frank H. K., 72, 74
Gruhn, John G., 23
Guy's Hospital, London, 1, 20
Guy's Hospital Reports, 1

Haddow, Alexander, 74
Hammersmith Hospital, 60, 82, 104,
    106
Hansen, Heine, 138, 143
Harley Street, 60
Harrison, Kent, 110
Hartley, Herman Otto, 60
Harvard University, 158
Heffernan, Joan, 107
Henschke, Claudia, 149, 151, 167–8
Higgins, Ian, 70
Hill, Austin Bradford, 72–5, 77–9, 80,
    102, 104, 105
Hilton, Gwen, 108, 116
Hilton, Matthew *Smoking in British
    Popular Culture* (2000), 64
Himsworth, Harold, 75

Histological Typing of Lung Tumours
    (1967), 140
histology, 7, 13–14, 108–9
historiography of medicine, 8
History of Cardiothoracic Surgery
    (Raymond Hurt 1996), 47, 49
History of Thoracic Surgery (Richard
    Meade 1961), 35, 46
HIV-AIDS, 148, 156
Holmes Sellors, Thomas, 121, 162
Hôspital Necker, 14
Hughes, Henry Marshall, 1–2, 20
Hunter, Robert, 102, 103, 105
Hurt, Raymond *History of
    Cardiothoracic Surgery* (1996), 47,
    49

Illich, Ivan, 175
Imperial Cancer Research Fund
    (ICRF), 104, 133, 144, 157
Infectious Disease (Notification) Act
    (1889), 31
influenza
    complications, 42
    epidemic (1918), 42
    see also empyema
innovations
    leading to new understandings of
        disease, 14, 22–6, 29–31
    practices, 14, 22–6, 29–31
    in surgery, 40
    technical, 14, 22–6, 34, 40
Institut Gustave Roussy, 140
Institute of Actuaries, 29
Institute of Cancer Research, London,
    60
International Association for the
    Study of Lung Cancer (IASLC),
    135, 142–3, 168
    International Staging Committee,
        143
    Lung Cancer (journal), 143
    Textbook of Lung Cancer, 143
    World Conferences on Lung
        Cancer, 142–3
International Commission on
    Radiological Protection, 74
International Journal of Epidemiology,
    76

International Reference Centre for the Histological Definition and Classification of Lung Tumours, 140
International Union against Cancer (Union Internationale Contre le Cancer, UICC), 139
  XIth International Cancer Congress (1974), 142
  Committee on Tumour Nomenclature and Statistics, 140
  TNM Committee, 141
International Workshop for the Therapy of Lung Cancer (1972), 142
intubation, 40, 41

Jachymov (St Joachimsthal), 91
Jackson Laboratory, Maine, 76
Jeger, Lena, 89, 90
Johns Hopkins Hospital, 49
Johns Hopkins Lung Project, 130
Joint National Cancer Survey Committee (Marie Curie Memorial and Queen's Institute of District Nursing), 67
Joules, Horace, 79–80, 91
Journal of the American Medical Association, 155
Journal of Thoracic Oncology, 143
Journal of Thoracic Surgery, 54

Kaplan, Henry, 135
Karnofsky, David, 62
Kavanagh, Terry, 166
Keating, Peter, 119, 132
Kemp, N. H., 145
Kennaway, Ernest, 66–7, 72–4, 91
Kennaway, Nina, 66–7, 74, 91
Keswick Jencks, Maggie, 163
King Edward VII Memorial Chest Hospital, 110
King Edward VII Welsh National Association, 40
King, George, 29–31
Kingsford, Edward, 1
Kissen, David, 152–3

Kmietowicz, Zosia, 149–50
Kreyberg, Leiv, 140

Laennec, R. T. H. (René Théophile Hyacinthe), 2, 14–16
  translation of his *traité de l'auscultation* into English, 16–17 *see also* Forbes, John
Laing, A. H., 134
Lancet, The (medical journal), 17
Lawrence, Nicola, 162–4
leeches, 21
lesions
  as specific markers of disease, 14–16, 24
  visibility in routine radiographs, 115
leukaemia, 7, 8
Leukaemia Research Fund, 144
Lewis, Thomas, 59
Lickint, Fritz, 76
Ligget group, 168
Lilienthal, Howard, 44–5
  operating bronchiectasis patients, 44–5
Lilly Oncology UK, 138
Little, Clarence, 76
Littlewood's Pools, 160
Liverpool
  City Council, 159, 160
  Daily Post, 159
  Echo, 159
  economic decline and social problems, 161
  Football Club, 161
  image and identity, 161
  Lung Project, 166–7, 169
  Philharmonic Orchestra, 160
  University, 159, 160, 166, 170
Llanbedr Hall, North Wales, 40
lobectomy, 35, 39, 42, 50, 97, 110
  performed by Davies, 39, 45–6
  performed by Lilienthal, 44–5
Logan, Andrew, 47
London Chest Hospital, 37, 39, 116
London School of Hygiene and Tropical Medicine (LSHTM), 72, 74, 107
Louis, Pierre Charles-Alexandre, 29
Löwy, Ilana, 157

Lundberg, George, 155
Lung Cancer (journal), 143
lung cancer
  cases recorded in the nineteenth
      century, 11, 13, 18–22, 24
  causes, 2, 73–5, 78, 80, 82, 90–1
    see also smoking
  and charity, 156–70
  and chemotherapy, 99, 106–7, 111,
      120, 121, 132–9, 144
  cell types, 120
  classification, 12, 14, 22, 24–5, 28,
      60, 108–9, 139–43
  clinical trials, 101–11, 132–9, 144–5
  comparisons
    with breast cancer, 26, 98, 126,
        144, 173–4
    with leukaemia, 132, 134–5
    with poliomyelitis, 80–1
    with prostate cancer, 67
    with stomach cancer, 98
  death rates, 67–72
    Figure 4.1, 68
    Figure 4.2, 69
  diagnosis, 1–2, 12, 13, 14, 18–22,
      25, 26–8, 32, 45, 46, 48, 96–7,
      115, 119–20
    see also microscopy; physical
        examination; sputum
        examination
  disease identity, 12, 16, 18, 22,
      171
  documentary film (1943), 61–2, 98
  and gender, 66, 67–8, 72
  increase in incidence, 2, 30, 31–2,
      65–7
  joint clinic, Royal Cancer (from
      1954 Marsden) and Brompton
      Hospitals, 53, 59, 96–7, 99–101
  first monograph published in
      English, 11
    see also Adler, Isaac
  non-small cell, 138
  and occupation, 66, 70, 91
  and radiotherapy, 53, 106–11, 116,
      120, 121, 134, 144
    see also radiotherapy
  rarity before the twentieth century,
      6, 11–12, 28, 31–2

  and other respiratory illnesses,
      14–16, 18–20, 25, 32–3, 46,
      67–71, 96, 97, 119, 147, 172
    Figure 4.1, 68
    Figure 4.2, 69
  and specialization, 139
  screening, 112–13, 126, 128–32,
      167–8
    Cooperative Early Lung Cancer
        Detection Project, 130–1
    cost, 129, 167–8
    Early Lung Cancer Action Project
        (ELCAP), 167–8
    paradoxical results, 131, 168
  small cell (or oat cell), 3, 25,
      109–11, 120, 139
  squamous cell, 48, 131
  staging, 7, 139–43
    effects on clinical practice, 141–2
    see also International Association
        for the Study of Lung Cancer
  and surgery, 3, 5, 6, 32–3, 34, 45–9,
      61–2, 98, 106–11, 113–17, 120,
      134, 144
    impact of computed tomography,
        141–2
    see also chest surgery; surgery
  survival rates 4, 97–8, 101, 112–13,
      116–17, 138
  symptoms when untreated, 1, 5,
      18–20, 25, 100–1
  terminal, 121–2
  treatability, 98
  treatment outcomes, 62, 97–8, 101,
      109, 112–13, 116–17, 120–1
    see also International Association
        for the Study of Lung Cancer
Lung Cancer Fund, 158–70, 173
lymph nodes, 142

Mackenzie, Basil William Sholto
      (second Baron Amulree), 74, 98
Macleod, Iain, 80–1
  press conference, 81, 85
Macmillan Cancer Care, 123, 157,
      166
Maggie's Cancer Care Centres, 163,
      165–6
magic bullets, 175

Magill, Ivan, 40–1
  Magill tube, 48
mammography, 127–8
Manchester Royal Infirmary, 65–6
Manhattan Cancer Society, 76
Marie Curie Hospital, 123
Marie Curie Memorial, Cancer Care,
  67, 123, 157, 166
Markle, Gerald, 155
Marlboro men, 125
Martin, F. M., 87–90
Mason, George, 47
mass radiography screening, 61,
  112–13, 129–30
  as a possible cause of lung cancer,
  72
  scaling down of services, 130, 131
Maudsley Personality Inventory
  (MPI), 153
Maulitz, Russell, 12
Maxwell, James, 74
Mayo Clinic, 60, 141, 142
  Mayo Lung Project, 130–1, 167
McCarthy, Peggy, 149, 151
McDowell, L. A., 112
McMichael, John, 60
Meade, Richard *History of Thoracic
  Surgery* (1961), 35, 46
medical journals
  and case histories, 13, 24
  and medical reform, 17
  and pathological anatomy, 16–17
medical oncology *see* oncology
Medical Research Council (MRC), 7,
  60, 70, 71–8, 144, 165, 174
  Cancer of the Lung Committee, 75
  Cancer Therapy Committee, 133,
  143
  Cancer Trials Office (CTO), 136
  clinical cancer research, 132–8
  Clinical Oncology and
    Radiotherapeutics Unit, 135
  Clinical Trials Unit, 136
  Committee on the Evaluation of
    Different Methods of Cancer
    Therapy, 101, 132–3
  Conference on the Evaluation of
    Different Methods of Cancer
    Therapy, 101

  Co-ordinating Committee on
    Cancer, 133
  Epidemiological Research Unit, 154
  Headquarters, 72, 74, 132, 134, 144
  lung cancer trials, 101–11
  Lung Cancer Working Group, 134,
  136
  lung cancer working party, 101–11,
  133, 134, 136
  Pneumoconiosis Research Unit, 70,
  82
  Principal Medical Officer, 74
  Radiotherapy Working Party, 132
  Social Medicine Research Unit, 73,
  125
  Statement on Tobacco Smoking and
    Cancer of the Lung (1957), 82,
  124
  Statistical Research Unit, 72, 73, 74
  Statistical Research and Services
    Unit, 134
  Tuberculosis (and Chest Diseases)
    Research Unit, 100, 107, 111,
  134, 136
  working parties on cancer
    treatment trials, 102–3
medullary tumours, 15–16
medulloblastoma, 102–3
Mellanby, Edward, 71
Meltzer, Samuel, 40, 54
Memorial Cancer Hospital, New York,
  62
Memorial Sloan Kettering Lung
  Project, 130
Meyer, Willy, 53–4
microscopy, 22–6
Middlesex Hospital, London, 17, 59,
  60, 73, 79, 101, 104, 110, 124–5,
  135
Midland Hotel, Manchester, 53
Mikulicz-Radecki, Johannes von, 38
  Negative Pressure Operating
    Chamber, 37–8 *Figure 3.1*
Miners' Disease, 91
Ministry of Health, 71, 74, 79, 81, 82,
  86, 98
Mitchell, Joseph, 102, 105, 135
molecular biology, 157
Moores, John, 160

morbidity data, 30–1
Morris, Eric, 158–9
mortality data, 30–1
Mountain, Clifton F., 141–3
Mueller, C. Barber, 49
Mukherjee, Siddhartha *The Emperor of
all Maladies* (2010), 8
Müller, Franz, 76
Müller, Johannes, 23

National Cancer Institute (NCI, US),
130, 141, 142, 150, 168
National Cancer Research Institute
(NCRI, UK), 149
National Health Service (NHS), 35, 56,
58
National Thoracic Surgery Service,
54–6
Natural History of Cancer (W. Roger
Williams 1908), 31
Neale, Julie, 133
neoplasms, 16
benign and malignant, 16
Newsholme, Arthur, 29
New York Association for Thoracic
Surgery, 53
Nicholson, Frank, 58
Nightingale, Florence, 29
nitrogen mustard, 62
nosology, 12
Nosworthy, Michael, 40

oat cell carcinoma *see* lung cancer,
small cell (or oat cell)
carcinoma
Ochsner, Alton, 75
oncology, 135
medical, 132, 137–8
thoracic, 139
*One in Three: A Son's Journey into the
History and Science of Cancer*
(Adam Wishart 2006), 8
O'Shaughnessy, Laurence, 47
Overholt, Richard, 75
Oxford University
Department of Social and
Preventive Medicine, 74

pain, 101
palliation 2, 3, 7, 21–2, 32, 100–1,
105, 121, 144
see also radiotherapy
palliative care, 121–3, 157
Palmer, J. W., 154
Papanicolaou (Pap) smear test, 126–7
Paris hospital medicine, 14, 17
Parker-Pope, Tara, 150
Pässler, Hans, 11
Paterson, Ralston, 59, 107
pathological anatomy, 12
availability of cadavers in post-
revolutionary Paris, 17
and cellular pathology, 23–6
and medical reform, 16–18
pathology, 51, 140
department at the Royal Cancer
Hospital, 59
patients
experiences, 25
see also Benbow, Mary; Castle,
Roy; Craig, Frank; Davies,
Anne; George VI
role during surgery, 52–3
support groups, 166
Pearson, Karl, 71
Pecek, Libor, 159
percussion (diagnostic procedure), 1,
13, 18, 19, 21
see also auscultation; physical
examination
Peters, Matthew, 151
Peto, Julian, 145
phthisis, 2, 6, 19, 21
cancerous, 15
physical examination, 1–2, 18–20
see also auscultation; percussion
Pickering, George, 60
Pilkington, Mavis, 160
Platt, Robert, 82–3
pleurisy, 21
pneumonectomy, 35, 48–9, 50, 97, 110
dissection pneumonectomy, 49
*On the Technique of Pneumonectomy
in Man* (Clarence Crafoord
1938), 49
performed by Evarts Graham, 48–9
pneumonia, 16, 18, 19, 21

pneumothorax problem, 37, 43, 52
pollution, atmospheric
    as a suspected cause of lung cancer,
        72–5, 91
    investigation by E. Kennaway, 73
    as a suspected cause of respiratory
        illness, 70
Poor Law, 30
Posner, E., 112
post mortem examinations, 2, 6, 12,
    13, 19, 22, 65–6, 99
    in British medicine, 17
    and microscopy, 24 *Figure 2.2*
    specificity of lung lesions, 14–16
Powell, Enoch, 84
Price Thomas, Clement, 88, 94–6, 100
*Primary Malignant Growths of the Lung
    and the Bronchi* (Isaac Adler 1912),
    11, 31–3
Primrose, A. F., 61
Pritchard, Colin, 154–5
private practice, 60
*Proceedings of the Royal Society of
    London*, 29–30
Procter, Robert, 76–7
progress
    attitudes towards, 8–9, 35
    narratives, 8–9, 35
    scientific and technological, 8–9
    see also innovations
public health policy, 123
    see also smoking, policy response

quality of life, 106–7, 134
Queen's Hospital, Sidcup, 40
Queen's Institute of District Nursing, 67

radiology (diagnostic), 50–1, 59, 97,
    115, 127, 141–2
radiotherapy, 3, 7, 59–61, 98
    Chair, London University, 60
    in combination with surgery, 49
    palliative, 7, 100–1
    presence in the MRC's Cancer
        Therapy Committee, 134
    radical, 98, 101, 106
    and randomized clinical trials,
        107–8
    status of, 59–61, 102, 135

Radium Commission, 71, 74
    Statistical Committee, 71, 74
randomized controlled trials (RCT),
    100, 103–4, 108–11, 119, 136,
    168
    ethics, 104–5
    recalcitrance, 4–9, 32–3, 111, 115,
        133, 173
    writing about, 9
regionalization, 35
Registrar General, 30
    Annual Report, 30
Reid, Hugh, 56, 158
Remak, Robert, 23
respiration
    manually assisted, 37, 40
    mechanically assisted, 37, 41
respiratory illness, 16
    history of, 12
risk, 64, 152, 169
Roberts, James Ernest Helme, 50–3, 99
Robinson, Kenneth, 81, 84–5
Robinson, Samuel, 42–3
Robson, Kenneth, 99
Rock Carling, Ernest, 74, 80
Rokitansky, Carl von, 16, 23–4
Rosenheim, Max, 133
Rouvray, Frederick George, 51
Rowbotham, Stanley, 41
Royal Cancer Hospital (from 1954:
    Royal Marsden), 53, 59–61, 66,
    74, 96–101
Royal College of Physicians (RCP)
    press conference, 82–3
    Report on Smoking and Health (1962),
        82–3, 84, 85, 100, 124
    Smoking and Health Now (1971), 85
Royal College of Surgeons of
    Edinburgh, 158
Rules for Radicals (activist text), 125
Russell, Dorothy, 103

saline purgatives, 1
sanitarians, 29
sarcoma, distinction from carcinoma,
    25
Sauerbruch, Ferdinand, 37, 53
    Negative Pressure Operating
        Chamber, 37–8 *Figure 3.1*

Saunders, Cicely, 123, 166
Scadding, John Guyett, 60, 99–100,
    103–4, 106–7, 111, 119, 133, 134,
    136
Scarff, Robert, 102, 105
Schairer, Eberhard, 76–7
Schleiden, Matthias, 23
Schneeberg, 91
Schöniger, Erich, 76–7
Schwann, Theodor, 23
scirrhus, 15, 16, 18, 26
screening
    breast cancer, 126, 127–8
        Cochrane Collaboration review,
            128
        Edinburgh study, 128
        Health Insurance Plan (HIP)
            study, 127–8
        NHS Breast Screening
            Programme, 128
    cervical cancer, 126–7, 166
    lung cancer *see* lung cancer,
        screening
    tuberculosis, 112–13, 128–30
second-hand smoke, 155–6
Second World War, 60
secondary prevention, 126
    see also early detection; screening
Seddon, Herbert, 103
sexually transmitted diseases, 148
Sheppard, David, 161–2
Simon, G., 115
Sitsen, A. E., 66
Sloan Kettering Institute, 62
small cell lung cancer *see* lung cancer,
    small cell
Smith, George Davey, 77
Smithers, David, 59–61, 71, 97,
    98–101, 119
smog *see* pollution
smoking, 64–92, 147–56
    addiction, 81
    attitudes towards, 78–9, 86–90,
        91–2, 123–6
    bans, 83–4, 88, 156
    campaigns by local councils,
        83–4
        Edinburgh campaign (1958),
            86–90

cause of lung cancer, 2–3, 6, 72–8
    investigation by Doll and Hill,
        72–5, 77–8
    investigation by Graham and
        Wynder, 75–6
    laboratory experiments, 78
    study by Hammond, 77–8
changing status in the 1970s,
    123–6, 153–4
and delinquency, 154–5
denormalization, 155
difference from other public health
    issues, 84
'excessive', 80, 81, 88
and gender, 85–6, 90, 153
and health education, 86–90, 119
in Nazi Germany, 76–7
passive, 155–6, 163
and personality, 152–3
policy response, 9, 77, 78–85, 91–2,
    123–6
and the press, 80–1, 124
prevalence in 1950s Britain, 78–9
as a risk factor, 64, 88–9, 152
and school children, 82, 89, 154–5
and social class, 86, 90, 153–4
television documentaries, 125
Smoking and Health (1964) *see* Surgeon
    General
Smoking and Health Now (1971) *see*
    Royal College of Physicians
Smoking and Health, Report on (1962)
    see Royal College of Physicians
Smoking in British Popular Culture
    (Matthew Hilton 2000), 64
Smyth, John, 144
Snow, John, 29
Snowdon, Christopher, 155
Socialist Medical Association, 79
Society for Experimental Biology and
    Medicine, 54
Society of Thoracic Surgeons (of Great
    Britain and Ireland), 35, 53–6, 114
soldering iron, use in chest surgery,
    47
South Manchester Hospital
    Management Committee, 58
South Wales and Montmouthshire
    Radiotherapy Service, 134

Southern General Hospital (Glasgow), 153
specialization, 35, 50
specificity of disease, 14
Spiro, Stephen, 145
sputum examination, 26
St Bartholomew's Hospital, 74, 133, 137, 144
St Christopher's Hospice, 122
St George's Hospital, 17
St Joachimsthal (Jachymov), 91
St Mary's Hospital Medical School, 158
St Thomas's Hospital, 40
staging, 140–1
    see also lung cancer; Tumour, Node, Metastasis (TNM) system
Standing Committee on Cancer and Radiotherapy of the Central Health Services Council, 74, 79–80
    Working Party on the Prevention of Cancer, 129–30
Stanford University, 135
statistics, 29–31
    methods of data collection 29, 30
    use in medicine and public health 29–31
stethoscope, 1, 14
    Stethoscope and Lungs, 15 *Figure 2.1*
Stewart, Alice, 74
stigma, 6, 90, 125–6, 147–56, 157, 164, 168, 170, 174
Stigma (Erving Goffman, 1968), 147
Stocks, Percy, 71–4, 80
Stoker, Michael, 144
Stokes, William, 18–20
Streptomycin, 100, 103
Stuber, Jennifer, 156
superior vena cava obstruction, 7, 18–19
Surgeon General, 42
    Smoking and Health, report (1964), 83
surgery
    gender connotations, 35
    heroic image, 35, 50
    open-and-close rate, 141

risks for surgeons, 39
    see also chest surgery; lung cancer and surgery
'Surgery in Chest Disease' (documentary film, 1943), 61–2
Sutherland, Ian, 134
Symposium on the Diagnosis and Treatment of Carcinoma of the Bronchus (Midhurst 1966), 113–15
Szabo, Jason, 8

Tall, Ruth, 134
tamoxifen, 167
Taylor, John, 20–2
Taylor, Peter, 6, 81, 94, 125
teamwork, 35, 52–3, 58–61, 62, 99
temperance movement, 155
Thatcher, Nick, 137–8, 145, 149
Thomson, J. G., 87–90
thoracic surgery *see* chest surgery
thoracoplasty, 51
thoracotomy, 46
Thorax (journal), 100
TNM Classification of Malignant Tumours (1968), 141
tobacco
    advertising, 84
    duties, 65, 81, 85, 89–90
    manufacturers, 83, 84, 124
        grant offer to MRC, 81
        Standing Committee, 80
Tobacco Settlement Fund (New York), 168
Tod, Margaret, 98
Todd, G. F., 80, 153–4
Todd, Ian, 134
tomography, 94
traditional remedies, 1, 5, 21
Trotter, Wilfred, 39
Troyer, Ronald, 155
Tubercle (journal), 100
tuberculosis, 2, 6, 12, 16, 31, 34, 67–8
    diagnostic technologies, 50–1
    sanatoria, 36, 40, 57
    services, 55
    surgical treatment of, 36
Tumour, Node, Metastasis (TNM) system, 140–1

tumours 14–16, 18, 19, 20, 22, 25–6,
    100
  of the brain, 135
  classification, 139–40
  grading, 108–9
  staging, *see* lung cancer; Tumour,
    Node, Metastasis (TNM) system
  see also accidental production,
    encephaloid, neoplasm,
    scirrhus

*Ueber den feineren Bau und die Formen
    der krankhaften Geschwülste*
    (Johannes Müller 1838), 23
Union Internationale Contre le
    Cancer (UICC) *see* International
    Union against Cancer
University College Hospital, London,
    5, 20, 22, 45, 59, 108, 116

Vaughan-Morgan, John, 82
Virchow, Rudolf, 23
Viscusi, W. Kip, 152

Walshe, Walter Hayle, 22
war
  casualties with wounds to face and
    jaws, 40–1
  see also First World War; Second
    World War
Ware, Martin, 74
Watson, Deirdre, 145
Welsh National School of Medicine,
    134
Wheeler-Bennett, John, 94–6
Whitehouse, J. N., 145

Wiernik, George, 134
Wilkes, Eric, 121–2
Williams, W. Roger *Natural History of
    Cancer* (1908), 31
Wilson, Harold, 84, 130
Windeyer, Brian, 59, 101–8, 132–3,
    135
Winston Man, 148
Wishart, Adam *One in Three: A Son's
    Journey into the History and Science
    of Cancer* (2006), 8
Witts, Lesley, 102, 103
World Conferences on Lung Cancer
    see International Association for
    the Study of Lung Cancer
World Health Assembly, Tenth
    (1957), 140
World Health Organization (WHO),
    139–40
  Expert Committee on Health
    Statistics, 139–40
  International Classification of
    Diseases, 140
Worlock, Derek, 161–2
Wykeham Brooks, William Donald,
    99
Wynder, Ernst, 75–6, 77–8
Wythenshawe Hospital, Manchester,
    57–8, 138

X-rays, 39
  see also radiology; radiotherapy

Yankelevitz, David, 167–8
Yarnold, John, 145
Years of Life Lost (YLL), 150

Printed and bound in the United States of America